Test-Driven Development with C++

A simple guide to writing bug-free Agile code

Abdul Wahid Tanner

BIRMINGHAM—MUMBAI

Test-Driven Development with C++

Copyright © 2022 Packt Publishing

Group Product Manager: Gebin George

Publishing Product Manager: Kunal Sawant

Senior Editor: Nithya Sadanandan

Technical Editor: Jubit Pincy

Copy Editor: Safis Editing

Project Coordinator: Deeksha Thakkar

Proofreader: Safis Editing

Indexer: Rekha Nair

Production Designer: Shyam Sundar Korumilli

Developer Relations Marketing Executive: Sonakshi Bubbar

First published: October 2022

Production reference: 1211022

Published by Packt Publishing Ltd.

Livery Place

35 Livery Street

Birmingham

B3 2PB, UK.

ISBN 978-1-80324-200-2

www.packt.com

Focus on constant improvement

– Abdul Wahid Tanner

Contributors

About the author

I'm **Abdul Wahid Tanner** and have been programming C++ since 1992 and teaching others how to code for almost as long. I first became interested in TDD around 2005 and decided to implement a board game that I liked playing using strict TDD. At the time, I was using C# and a test framework that was able to find and run the tests. It wasn't until about 2014 that I decided to see whether I could use C++ and TDD together. There were many choices for tools that helped run C++ tests, but they were big and full of features. I wanted something small and simple, so I wrote a unit test library in about 650 lines of code. The code in this book is a complete and better rewrite of that early code. I hope you find the code in this book useful and decide to use it in your projects. However, even if you decide to use one of the commercial tools, you should still benefit from the understanding that this book provides about how the tools work and how to follow the TDD process.

About the reviewer

Serban Stoenescu is a software developer specializing in C++. He is currently working for Everseen on some projects that require OpenCV and some machine learning. In the past he has worked for SkAD Labs on a CAD project with OpenCASCADE, Atigeo on Big Data projects in Java, and Alcatel-Lucent (now part of Nokia). At Alcatel-Lucent, he was also a trainer, teaching Rational ROSE. He is the co-author of "Towards the Impact of Design Flaws on the Resources Used by an Application", a paper at ARMS-CC 2014. Currently, he is also a Udemy instructor, his most popular course being "C++ Unit Testing: Google Test and Google Mock". His other courses are "CMake from Zero to Hero" and "C++ Machine Learning Algorithms Inspired by Nature".

Table of Contents

Preface   xiii

Part 1: Testing MVP

1

Desired Test Declaration   3

Technical requirements	3	How can we use C++ to write tests?	11
What do we want tests to do for us?	4	How will the first test be used?	15
What should a test look like?	5	Summary	18
What information does a test need?	6		

2

Test Results   19

Technical requirements	19	Summarizing the results	28
Reporting a single test result	20	Redirecting the output results	31
Enhancing the test declaration to support multiple tests	25	Summary	33

3

The TDD Process   35

Technical requirements	35	Enhancing a test and getting another pass	43
Build failures come first	36		
Do only what is needed to pass	38	Summary	57

4

Adding Tests to a Project   59

Technical requirements	60	Enhancing the testing library to support assertions	67
How to detect whether a test passes or fails	60	Should error cases be tested, too?	72
		Summary	74

5

Adding More Confirm Types   75

Technical requirements	75	Confirming string literals	94
Fixing the bool confirms	76	Confirming floating point values	96
Confirming equality	76	How to write confirms	101
Decoupling test failures from line numbers	82	Summary	103
Adding more confirm types	87		

6

Explore Improvements Early   105

Technical requirements	105	Exploring lambdas for tests	108
Getting line numbers without macros	106	Summary	113

7

Test Setup and Teardown   115

Technical requirements	116	Handling errors in setup and teardown	145
Supporting test setup and teardown	116	Summary	150
Enhancing test setup and teardown for multiple tests	123		

8

What Makes a Good Test?   151

Technical requirements	151	Only test your project	159
Making tests easy to understand	152	Test what should happen instead of how	159
Keeping tests focused on specific scenarios	154	Summary	160
Use random behavior only in this way	157		

Part 2: Using TDD to Create a Logging Library

9

Using Tests   163

Technical requirements	163	Logging and confirming the first message	172
Why build a logging library?	164	Adding timestamps	179
How will TDD help build a logging library?	165	Constructing log messages with streams	181
What would the ideal logging library look like?	166	Summary	185
Starting a project using TDD	169		

10

The TDD Process in Depth   187

Technical requirements	188	Controlling what gets logged	235
Finding gaps in the testing	188	Enhancing filtering for relative matches	245
Adding log levels	190	When is testing too much?	261
Adding default tag values	199	How intrusive should tests be?	263
Exploring filtering options	212	Where do integration or system tests go in TDD?	264
Adding new tag types	216	What about other types of tests?	265
Refactoring the tag design with TDD	221	Summary	266
Designing tests to filter log messages	226		

11

Managing Dependencies   269

Technical requirements	269	Adding multiple logging outputs	274
Designing with dependencies	270	Summary	284

Part 3: Extending the TDD Library to Support the Growing Needs of the Logging Library

12

Creating Better Test Confirmations   287

Technical requirements	288	Enhancing the test library to support Hamcrest matchers	294
The problem with the current confirmations	288	Adding more Hamcrest types	300
Simplifying string confirmations	290	Summary	308

13

How to Test Floating-Point and Custom Values   309

Technical requirements	310	matchers	322
More precise floating-point comparisons	310	Writing custom Hamcrest matchers	327
Adding floating-point Hamcrest		Summary	333

14

How to Test Services   335

Technical requirements	335	Introducing the SimpleService project	340
Service testing challenges	336	Summary	350
What can be tested in a service?	338		

15

How to Test With Multiple Threads   351

Technical requirements   352
Using multiple threads in tests   352
Making the logging library
thread-safe   361
The need to justify multiple threads   367
Changing the service return type   375

Making multiple service calls   379
How to test multiple threads
without sleep   385
Fixing one last problem detected
with logging   398
Summary   400

Index   401

Index   401

Other Books You May Enjoy   406

Other Books You May Enjoy   406

Preface

Many years ago, I started avoiding cookbook-style books when I wanted to learn a new technology or develop a new skill. A book should guide us to understand fundamental ideas so we can adapt our learning to fit our needs. We should not get lost just because a recipe doesn't match exactly what we want to accomplish.

I also tend to avoid books that are full of theory. You probably know what I mean. Instead of explaining how to use a technology, these books mention ideas with no exploration of putting the ideas into practice. Focusing on either one of these two tangents can leave the learning experience incomplete.

This book is different. Instead of giving you a few simple examples like a cookbook or droning on with rules and definitions, I'll show you how to use tests to write software that does the following:

- What users expect and with ease
- Let's you add new features without breaking what you've already done
- Clearly communicates what the code does

We're going to start from the beginning and build multiple projects. You'll learn what a test is and how it can help by building an easy and natural way to write tests in C++. This is called **Test-Driven Development (TDD)** and you'll be using TDD to develop the tools needed for your projects.

In order to get the most benefit from TDD, you'll need a way to write and run tests. Some programming languages already support TDD while C++ needs extra tools and libraries to support tests. I could have simply explained how to use various TDD tools that already exist. That would be a cookbook approach because each existing solution is slightly different. By starting from the beginning and building what you'll need, you'll learn TDD and at the same time you'll learn how all the other solutions work. The concepts are the same.

For example, most people who learn how to drive a car will be able to drive a pickup truck or a van. Some minor differences are expected but the concepts are the same. Riding a motorcycle is more of a challenge but many of the same concepts are still valid. Driving a bus might need some extra training but is also similar.

If you decide to use other tools or even other languages, you'll gain a deep understanding of how those other tools work by following the steps in this book. The other tools will be like driving a pickup truck.

This book is divided into three parts. You'll build a testing library in the first part that will let you write and run tests. You'll learn how to proceed one step at a time and let the tests guide you.

The second part will use the testing library to start a new project to help you log messages as your software runs. This is a logging library and this project will show you how to build software that meets the needs of your intended customer. You'll be adding many features to the logging library and you'll learn how to keep everything working while developing more features.

The intended customer for the logging library is a C++ micro-services developer and this customer focus extends into the third part where you'll extend the testing library and build a simple service. The service will be used to explain one of the most difficult aspects of software development how to test multiple threads.

All three parts will combine to show you how to use TDD to effectively design and write software that meets the customers needs, gives you the confidence to make changes, and lets other team members understand the design.

Who this book is for

This book is for C++ developers already familiar with and using C++ for daily tasks. You don't need to be an expert, but you should already know some modern C++ and how to use templates to get the most out of this book.

If you've ever struggled to add a new feature to a large project without breaking the software in other places, then this book can help. You'll become a better software developer and avoid the stress and worry of fixing bugs or making design changes. TDD is a process that will guide you to create intuitive designs that your users will understand and enjoy.

If you've ever hesitated over making an improvement and decided to keep an outdated and confusing design that nobody wants to touch because it might break, then this book can help. Maybe you have a good idea but you know your manager will never agree because the risk is too high. If your project is following TDD practices that this book explains, then you can feel more confident about making improvements. You can let the tests prove your idea is solid.

Do you have new team members who need to quickly understand how your software works? The process that this book explores will help you write tests in TDD that will document the software design and requirements. You might even find that the tests help your own understanding especially when returning to a project that you haven't worked on in a long time.

C++ is a language with a lot of power and details to consider. This book helps to simplify the software development process through TDD. Even if you already know how to use one or more TDD tools, this book will improve your skills and give you the confidence to apply TDD in more advanced or larger projects.

What this book covers

Chapter 1, Desired Test Declaration, uses C++ to write your first test starting from an empty project.

Chapter 2, Test Results, reports and uses test results so that you can quickly know what parts of a project are working and what parts are failing.

Chapter 3, The TDD Process, uses TDD on purpose by identifying the steps you've already been doing.

Chapter 4, Adding Tests to a Project, enhances tests to pass or fail based on whether your expectations are being met.

Chapter 5, Adding More Confirm Types, makes tests more capable by giving you the power to check any value types to see if they meet expectations.

Chapter 6, Explore Improvements Early, looks for other improvements and shows you how to step back and explore ideas you might not have thought of up to this point.

Chapter 7, Test Setup and Teardown, prepares tests to run and the cleanup afterward so that you can focus tests on what they need to do which helps make the tests easier to understand.

Chapter 8, What Makes a Good Test?, provides advice and learning from other chapters to reinforce what you've learned so far and encourages you to continue with hints of topics in later chapters.

Chapter 9, Using Tests, builds a logging library that puts everything you've learned so far into practical use.

Chapter 10, The TDD Process in Depth, practices using TDD to add features to a growing project including how to handle design changes.

Chapter 11, Managing Dependencies, adds features that can be exchanged, showing you how to test unreliable software components and how to make progress even when you don't have all the components that you need.

Chapter 12, Creating Better Test Confirmations, uses Hamcrest matchers to improve tests by letting you verify expectations more naturally.

Chapter 13, How To Test Floating Point and Custom Values, shows how to reliably use floats and how to extend Hamcrest matchers to meet custom needs.

Chapter 14, How To Test Services, covers how services are different and how to test services.

Chapter 15, How To Test With Multiple Threads, simplifies and coordinates tests with multiple threads so that you can avoid race conditions while testing every possible thread interaction reliably and predictably.

To get the most out of this book

You will need a modern C++ compiler capable of building C++20 or later code. Everything in this book uses standard C++ and will run on any computer. All output is text.

This book lets you build the code with whatever build system you are most comfortable using. The projects all have few files with simple folder structures. You can easily create a project in your code editor and follow along.

This book does not describe finished projects. Each project is a journey. The first chapter starts with an empty project and each subsequent chapter adds to the code already created in previous chapters. You're encouraged to follow along.

Remember to study the process. This is more important than the actual code that each chapter explains. The TDD process is mentioned many times in this book. The first few chapters introduce you to TDD until there are enough features available in the testing library so that the testing library can be used to build another project. The second logging library project explores the TDD process in more depth and detail. Once the logging library is usable, then both the testing and logging libraries are used to build a simple service project. There is a pattern to the TDD process that repeats. Learning the TDD process through each of the projects will give you the most benefit.

Because the projects are built step by step, you can also benefit from the mistakes explained along the way. Sometimes, the designs will change, and you can benefit from the reasons for the changes as well as learn how to manage the changes.

When I say that you need a C++20 compiler, that's a simplification. There are many different features of C++ that compiler vendors support in different versions of their compilers. A good rule to follow is to make sure that your compiler supports *concepts* in C++20. We use concepts in the final chapters of this book and if your compiler supports concepts, then you should have everything you need. A good link to read more is:

```
https://en.cppreference.com/w/cpp/compiler_support
```

When you visit the link, scroll down to the C++20 section and look for the row that identifies the concepts feature. At the time of writing, the following compilers should work:

- GCC version 10

- Clang version 10

- MSVC version 19.30

- Apple Clang version 12 (however, this compiler only has a partial implementation)

You'll likely be able to use whatever version of the C++ Standard Library that comes with your compiler. However, another good rule that should work is to make sure that your Standard Library also supports concepts. At the time of writing, the following Standard Libraries should work:

- GCClibstdc++ version 10

- Clang libc++ version 13

- MSVC STL version 19.23

- Apple Clang version 13.1.6

If you are using the digital version of this book, we advise you to type the code yourself or access the code from the book's GitHub repository (a link is available in the next section). Doing so will help you avoid any potential errors related to the copying and pasting of code.

Download the example code files

You can download the example code files for this book from GitHub at `https://github.com/ PacktPublishing/Test-Driven-Development-with-CPP`. If there's an update to the code, it will be updated in the GitHub repository.

We also have other code bundles from our rich catalog of books and videos available at `https://github.com/PacktPublishing/`. Check them out!

Conventions used

There are a number of text conventions used throughout this book.

`Code in text`: Indicates code words in text, database table names, folder names, filenames, file extensions, pathnames, dummy URLs, user input, and Twitter handles. Here is an example: "That's why the confirm macros we have now are called `CONFIRM_TRUE` and `CONFIRM_FALSE`."

A block of code is set as follows:

```
TEST("Test bool confirms")
{
    bool result = isNegative(0);
    CONFIRM_FALSE(result);

    result = isNegative(-1);
    CONFIRM_TRUE(result);
}
```

Any command-line input or output is written as follows:

```
Running 3 tests
- - - - - - - - - - - - - - -
Test can be created
Passed
- - - - - - - - - - - - - - -
Test with throw can be created
Passed
- - - - - - - - - - - - - - -
Test that never throws can be created
Failed
Expected exception type int was not thrown.
- - - - - - - - - - - - - - -
Tests passed: 2
Tests failed: 1
Program ended with exit code: 1
```

Bold: Indicates a new term, an important word, or words that you see onscreen. For instance, words in menus or dialog boxes appear in **bold**. Here is an example: "We'll just change the status from **Failed** to **Expected failure**."

> **Tips or important notes**
> Appear like this.

Get in touch

Feedback from our readers is always welcome.

General feedback: If you have questions about any aspect of this book, email us at customercare@ packtpub.com and mention the book title in the subject of your message.

Errata: Although we have taken every care to ensure the accuracy of our content, mistakes do happen. If you have found a mistake in this book, we would be grateful if you would report this to us. Please visit www.packtpub.com/support/errata and fill in the form.

Piracy: If you come across any illegal copies of our works in any form on the internet, we would be grateful if you would provide us with the location address or website name. Please contact us at copyright@packt.com with a link to the material.

If you are interested in becoming an author: If there is a topic that you have expertise in and you are interested in either writing or contributing to a book, please visit authors.packtpub.com.

Share Your Thoughts

Once you've read *Test-Driven Development with C++*, we'd love to hear your thoughts! Scan the QR code below to go straight to the Amazon review page for this book and share your feedback.

https://packt.link/r/1803242000

Your review is important to us and the tech community and will help us make sure we're delivering excellent quality content.

Download a free PDF copy of this book

Thanks for purchasing this book!

Do you like to read on the go but are unable to carry your print books everywhere? Is your eBook purchase not compatible with the device of your choice?

Don't worry, now with every Packt book you get a DRM-free PDF version of that book at no cost.

Read anywhere, any place, on any device. Search, copy, and paste code from your favorite technical books directly into your application.

The perks don't stop there, you can get exclusive access to discounts, newsletters, and great free content in your inbox daily

Follow these simple steps to get the benefits:

1. Scan the QR code or visit the link below

https://packt.link/free-ebook/9781803242002

2. Submit your proof of purchase
3. That's it! We'll send your free PDF and other benefits to your email directly

Part 1:
Testing MVP

This book is divided into three parts. In this first part, you'll learn why Test-Driven Development is important and how to use it to help you design and write software. We'll be starting with an empty project and using Test-Driven Development practices to design and build a unit test library. You'll learn everything you need by following along as we explore ideas and build a working project one step at a time.

The following chapter will be covered in this part:

- *Chapter 1, Desired Test Declaration*
- *Chapter 2, Test Results*
- *Chapter 3, The TDD process*
- *Chapter 4, Adding Tests to a Project*
- *Chapter 5, Adding More Confirm Types*
- *Chapter 6, Explore Improvements Early*
- *Chapter 7, Test Setup and Teardown*
- *Chapter 8, What Makes a Good Test?*

1

Desired Test Declaration

If we're going to have a **test-driven development** (**TDD**) process, we need tests. This chapter will explain what the tests will do, how we will write them, and how we will use them.

We'll be starting from the very beginning and slowly building a full library to help manage and run tests, and we'll be using the test library to help build itself. Initially, there will only be a single test. The following chapters will add more capabilities and grow the test library.

Starting with the end goal in mind, we'll first think about what it will be like to create and use a test. Writing tests is a big part of TDD, so it makes sense to start thinking about testing even before we have the ability to create and run tests.

TDD is a process that will help you design better code and then make changes to your code without breaking parts that you've already verified to be working as expected. In order for this process to work, we need to be able to write tests. This chapter will explore what tests can do for us and how we can write them.

We'll cover the following main topics in this chapter:

- What do we want tests to do for us?
- What should a test look like?
- What information does a test need?
- How can we use C++ to write tests?
- How will the first test be used?

Technical requirements

All the code in this chapter uses standard C++, which builds on any modern C++ 17 or later compiler and standard library. Future chapters will require C++ 20 but for now, only C++ 17 is needed. The number refers to the year that the standard was approved and finalized, so C++ 17 was released in 2017 and C++ 20 was released in 2020. Each release adds new features and capabilities to the language.

The code we'll be working with starts with an empty console project with a single source file called `main.cpp`.

If your development environment gives you a "Hello, world!" project when starting a new command line or console project, you can delete the contents of the `main.cpp` file because this chapter will start from the very beginning with an empty file.

You can find all the code for this chapter at the following GitHub repository: `https://github.com/PacktPublishing/Test-Driven-Development-with-CPP`.

What do we want tests to do for us?

Before we start learning about test-driven development, what it is, and what the process involves, let's step back and think about what we want. Without knowing all the details about what a test is, let's ask ourselves what our tests should look like.

I like to relate programming concepts to everyday experiences whenever possible. Maybe you have an idea to solve a problem that you have noticed and want to see whether your idea will work. If you wanted to test this idea before announcing it to the world, how would you do it?

You probably won't be able to test everything to do with your idea at once. What would that even mean? There are probably small parts of your idea that you can think about initially. These should be easier to test and should help to clarify your idea and get you to think of other things to test.

So, let's focus on simply testing small parts of the idea, whatever it is. You'd want to get everything set up and then start some actions or steps that should tell you whether each part works or not. Some of your tests might work well and some might cause you to rethink your idea. This is definitely better than jumping into the full idea without knowing whether it will work or not.

To put this into a real context, let's say you have an idea to build a better broom. That's a vague idea that's hard to envision. However, let's say that while sweeping the floor recently, you noticed your arms getting sore and thought that there had to be a better way. Thinking about the actual problem is a good way to turn a vague idea into something with a more solid meaning.

Now, you might start thinking about testing broom handles of different shapes, different grips, or different sweeping motions. These are the smaller parts of the idea that can be tested. You can take each grip or motion and turn it into a set of steps or actions that will test that part until you find one that works best.

Well, in programming, a set of steps can be a **function**. It doesn't matter what that function does right now. We can think of each test as represented by a function. If you can call a function and it gives you the expected result, then you can say that the test passed. We'll build on this idea throughout this book.

Now that we've decided to use a function for a test, what should it look like? After all, there are lots of ways to write a function.

What should a test look like?

It should be as simple to write a test as it is to declare and write a function, and we should be able to simplify things even further. A normal function can have whatever return type you want, a name, a set of parameters, and a body of code.

A function is also something that you write so that it can be called by other code. This code should know what the function does, what it returns, and what arguments need to be passed. We'll keep things simple for our test functions and only worry about the name for now.

We want each test function to have its own name. Otherwise, how would we be able to keep track of all the various tests we'll eventually be writing? As for the return type, we haven't identified an actual need yet, so we'll use void.

You'll learn more about this process in *Chapter 3, The TDD Process*. When using TDD, don't get ahead of yourself. Only do what you need to do at the time. As with the void return type, we'll also not have any parameters.

It might seem too simple but this is a good start. So far, a test is nothing more than a function, which returns nothing and takes no parameters. It has a name to identify it and will include whatever code is needed to run the test.

Because we're going to start using TDD to help design a simple testing library, our first test should ensure that we can create a test. This is a simple start, which defines a test function and calls it from main. All of this is in a single file called main.cpp:

```
#include <iostream>

void testCanBeCreated ()
{
    std::cout << "testCanBeCreated" << std::endl;
}

int main ()
{
    testCanBeCreated();

    return 0;
}
```

You might be thinking that this is not a test, it's just a function that prints its own name, and you'd be right. We're going to build it up from the very beginning in an agile manner using only what we have available. Right now, we don't have a test library to use yet.

Still, this is starting to resemble what we eventually want. We want a test to be just like writing a function. If you build and run the project now, the output is as expected:

```
testCanBeCreated
Program ended with exit code: 0
```

This shows the output from running the program. It displays the name of the function. The text in the second line actually comes from my development tools and shows the program exit code. The exit code is the value returned from main.

This is a start but it can be improved. The next section will look at what information a test needs, such as its name.

What information does a test need?

The current test function doesn't really know its name. We want the test to have a name so that it can be identified but does that name really need to be the name of the function? It would be better if the name was available as data so it could be displayed without hardcoding the name inside the test body.

Equally, the current test function doesn't have any idea of success or failure. We purposefully ignored the test result until now, but let's think about it. Is it enough for a test function to return the status? Maybe it needs a bool return type where true would mean success and false would mean the test failed.

That might be a bit too simplistic. Sure, it would be enough for now, but if a test fails, it might be good to know why. A bool return type won't be enough later. Instead of designing the entire solution, we just need to figure out what to do that will meet the expected needs.

Since we already know that we need some data to hold the test name, what if we now add simple bool result data in the same place? This would let us keep the test function return type as void, and it leaves room for a more advanced solution later.

Let's change the test function into a **functor** as follows so that we can add member data for the name and result. This new design moves away from using a simple function for a test. We need a **class** to hold the data for the name and result. A functor is a class that can be called like a function using operator(), as this code shows:

```
#include <iostream>
#include <string_view>

class Test
```

```
{
public:
    Test (std::string_view name)
    : mName(name), mResult(true)
    {}

    void operator () ()
    {
        std::cout << mName << std::endl;
    }

private:
    std::string mName;
    bool mResult;
};

Test test("testCanBeCreated");

int main ()
{
    test();

    return 0;
}
```

The biggest problem with this is that we no longer have a simple way to write a test as if it was a simple function. By providing operator (), or *function call operator*, we created a functor that will let us call the class as if it was a function from within the main function. However, it's more code to write. It solves the problem of the test name, gives us a simple solution for the result, which can be expanded later, and also solves another problem that wasn't obvious before.

When we called the test function in main before, we had to call it by the function name. That's how functions are called in code, right? This new design eliminates that coupling by creating an instance of the Test functor called test. Now, main doesn't care about the test name. It only refers to the instance of the functor. The only place in which the name of the test now appears in the code is when the functor instance is created.

We can fix the problem of all the extra code needed to write a test by using a *macro*. Macros are not needed in C++ as they used to be and some people might even think that they should be removed from the language entirely. They do have a couple of good uses left and *wrapping up code* into a macro is one of them.

We'll eventually put the macro definition into a separate header file, which will become the test library. What we want to do is wrap up all the functor code in the macro but leave the implementation of the actual test function body to be written as if everything was a normal function.

First, we'll make a simple change to move the implementation of the test function body outside of the class definition, like this. The function call operator is the method that needs to be moved outside:

```
class Test
{
public:
    Test (std::string_view name)
    : mName(name), mResult(true)
    {}

    void operator () ();

private:
    std::string mName;
    bool mResult;
};

Test test("testCanBeCreated");

void Test::operator () ()
{
    std::cout << mName << std::endl;
}
```

Then, the class definition, instance declaration, and first line of the function call operator can be turned into a macro. Compare the following code with the previous code to see how the `Test` class is turned into the `TEST` macro. By itself, this macro would not compile because it leaves the function call operator in an unfinished state. That's exactly what we want because it lets the code use the macro like a function signature declaration and finish it up by providing the code inside the curly braces:

```
#define TEST class Test \
{ \
public: \
    Test (std::string_view name) \
    : mName(name), mResult(true) \
    {} \
    void operator () (); \
private: \
    std::string mName; \
    bool mResult; \
}; \
Test test("testCanBeCreated"); \
void Test::operator () ()

TEST
{
    std::cout << mName << std::endl;
}
```

Because the macro is defined over multiple lines, each line except the last needs to end with a backslash. The macro is a little more compact because the empty lines have been removed. This is a personal choice and you can leave the empty lines if you want. An empty line still needs the backslash though, which defeats the purpose of having an empty line.

The code uses the TEST macro with the unfinished function call operator just like a function definition, but then it completes the code by providing the curly braces and method implementation needed.

We're making progress! It might be hard to see it because everything is in a single file. Let's fix that by creating a new file called Test.h and moving the macro definition to the new file, like this:

```
#ifndef TEST_H
#define TEST_H

#include <string_view>

#define TEST class Test \
{ \
public: \
    Test (std::string_view name) \
```

```
        : mName(name), mResult(true) \
        {} \
        void operator () (); \
private: \
        std::string mName; \
        bool mResult; \
}; \
Test test("testCanBeCreated"); \
void Test::operator () ()


#endif // TEST_H
```

Now, we can go back to simpler code in `main.cpp`, like this next block of code shows. All we need to do is include `Test.h` and we can use the macro:

```
#include "Test.h"

#include <iostream>

TEST
{
    std::cout << mName << std::endl;
}

int main ()
{
    test();

    return 0;
}
```

We now have something that's beginning to look like the simple function we started with, but there's a lot of code hidden inside the TEST macro to make it seem simple.

In the next section, we'll fix the need for `main` to call `test()` directly. The name of the functor, `test`, is a detail that should not be known outside of the macro, and we definitely shouldn't need to call a test directly to run it, no matter what it's called.

How can we use C++ to write tests?

Calling the test directly might not seem like a big problem right now because we only have one test. However, as more tests are added, the need to call each one from `main` will lead to problems. Do you really want to have to modify the `main` function every time you add or remove a test?

The C++ language doesn't have a way to add extra custom information to a function or a class that could be used to identify all the tests. So, there is no way to look through all the code, find all the tests automatically, and run them.

One of the tenets of C++ is to avoid adding language features that you might not need, especially language features that affect your code without your awareness. Other languages might let you do other things, such as adding custom attributes, which you can use to identify tests. C++ defines standard attributes, which are intended to help the compiler optimize code execution or improve the compilation of your code. The standard C++ attributes are not something that we can use to identify tests and custom attributes would go against the tenet of unneeded features. I like this about C++, even if it means that we have to work a little harder to figure out which tests to run.

All we need to do is let each test identify itself. This is different from writing code that would try to find the tests. Finding the tests requires that they be marked in some way, such as using an attribute, so that they stand out and this isn't possible in C++. Instead of *finding them*, we can use the constructor of each test functor so that they *register themselves*. The constructor for each test will add itself to the registry by pushing a pointer to itself onto a collection.

Once all the tests are registered through addition to a collection, we can go through the collection and run them all. We already simplified the tests so that they can all be run in the same way.

There's just one complication that we need to be careful about. The test instances that are created in the `TEST` macro are global variables and can be spread out over many different source files. Right now, we have a single test declared in a single `main.cpp` source file. We'll need to make sure that the collection that will eventually hold all the registered tests is set up and ready to hold the tests before we start trying to add tests to the collection. We'll use a function to help coordinate the setup. This is the `getTests` function, shown in the following code. The way `getTests` works is not obvious and is described in more detail after the next code.

Now is also a good time to start thinking about a **namespace** to put the testing library into. We need a name for the namespace. I thought about what qualities stand out in this testing library. Especially when learning something like TDD, simplicity seems important, as is avoiding extra features that might not be needed. I came up with the word *mere*. I like the definition of mere: *being nothing more nor better than*. So, we'll call the namespace `MereTDD`.

Here is the first part of the Test.h file with the new namespace and registration code added. We should also update the include guard to something more specific, such as MERETDD_TEST_H, like this:

```cpp
#ifndef MERETDD_TEST_H
#define MERETDD_TEST_H

#include <string_view>
#include <vector>

namespace MereTDD
{

class TestInterface
{
public:
    virtual ~TestInterface () = default;

    virtual void run () = 0;
};

std::vector<TestInterface *> & getTests ()
{
    static std::vector<TestInterface *> tests;

    return tests;
}

} // namespace MereTDD
```

Inside the namespace, there is a new TestInterface class declared with a run method. I decided to move away from a functor and to this new design because when we need to actually run the test later, it looks more intuitive and understandable to have a method called run.

The collection of tests is stored in a *vector* of TestInterface *pointers*. This is a good place to use raw pointers because there is no ownership implied. The collection will not be responsible for deleting these pointers. The vector is declared as a *static* variable inside the getTests function. This is to make sure that the vector is properly initialized, even if it is first accessed from another .cpp source file compilation unit.

C++ language makes sure that *global variables* are initialized before main begins. That means we have code in the test instance constructors that get run before main begins. When we have multiple .cpp files later, making sure that the collection is initialized first becomes important. If the collection is a normal global variable that is accessed directly from another compilation unit, then it could be that the collection is not yet ready when the test tries to push itself onto the collection. Nevertheless, by going through the getTests function, we avoid the readiness issue because the compiler will make sure to initialize the static vector the *first time that the function is called.*

We need to scope references to classes and functions declared inside the namespace anytime they are used within the macro. Here is the last part of Test.h, with changes to the macro to use the namespace:

```
#define TEST \
class Test : public MereTDD::TestInterface \
{ \
public: \
    Test (std::string_view name) \
    : mName(name), mResult(true) \
    { \
        MereTDD::getTests().push_back(this); \
    } \
    void run () override; \
private: \
    std::string mName; \
    bool mResult; \
}; \
Test test("testCanBeCreated"); \
void Test::run ()


#endif // MERETDD_TEST_H
```

The Test constructor now registers itself by calling getTests and pushing back a pointer to itself to the vector it gets. It doesn't matter which .cpp file is being compiled now. The collection of tests will be fully initialized once getTests returns the vector.

The TEST macro remains outside of the namespace because it doesn't get compiled here. It only gets inserted into other code whenever the macro is used. That's why inside the macro, it now needs to qualify TestInterface and the getTests call with the MereTDD namespace.

Inside `main.cpp`, the only change is how to call the test. We no longer refer to the test instance directly and now iterate through all the tests and call `run` for each one. This is the reason I decided to use a method called `run` instead of the function call operator:

```
int main ()
{
    for (auto * test: MereTDD::getTests())
    {
        test->run();
    }

    return 0;
}
```

We can simplify this even more. The code in `main` seems like it needs to know too much about how the tests are run. Let's create a new function called `runTests` to hold the `for` loop. We might later need to enhance the `for` loop and this seems like it should be internal to the test library. Here is what `main` should look like now:

```
int main ()
{
    MereTDD::runTests();

    return 0;
}
```

We can enable this change by adding the `runTests` function to `Test.h` inside the namespace, like this:

```
namespace MereTDD
{

class TestInterface
{
public:
    virtual ~TestInterface () = default;

    virtual void run () = 0;
```

```
};

std::vector<TestInterface *> & getTests ()
{
    static std::vector<TestInterface *> tests;

    return tests;
}

void runTests ()
{
    for (auto * test: getTests())
    {
        test->run();
    }
}

} // namespace MereTDD
```

After all these changes, we have a simplified `main` function that just calls on the test library to run all the tests. It doesn't know anything about which tests are run or how. Even though we still have a single test, we're creating a solid design that will support multiple tests.

The next section explains how you will use tests by looking at the first test.

How will the first test be used?

So far, we have a single test that outputs its name when run, and this test is declared inside of `main`. `cpp`. This is not how you'll want to declare your tests going forward. I've mentioned having multiple `.cpp` files with multiple tests in each one. We're not ready for that yet but we can at least move the single test that we have into its own `.cpp` file.

The whole point of declaring multiple tests in multiple `.cpp` files is to help organize your tests. Group them into something meaningful. We'll get to multiple tests later. For now, what is the purpose of our single test?

It is supposed to show that a test can be created. There may be other aspects of test creation that we'll be interested in. So, it might make sense to create a `.cpp` file focused on test creation. Inside this `.cpp` file would be all the tests relating to *different ways to create tests*.

You can organize your tests however you want. If you have a project you are working on that has its own set of source files, it might make sense to group your tests around the source files. So, you would have a test .cpp file with many tests inside, which are all designed to test everything related to a .cpp file from your actual project. This would make sense if your project files were already organized well.

Or, you might take a more functional approach to organizing your tests. Since we only have a single header file called Test.h that we need to test, instead of also creating a single .cpp file to hold all the tests, let's take a functional approach and split the tests based on their purpose.

Let's add a new .cpp file to the project called Creation.cpp and move the single test that we have so far into the new file. At the same time, let's think for a moment about how we will use the test library later on.

What we're building is not really a library that gets compiled and linked into another project. It's just a single header file called Test.h, which other projects can include. It's still a library, just one that gets compiled alongside the other project.

We can even start treating the tests we have now this way. In the project structure, we have Test.h and main.cpp so far. The main.cpp file is similar to that of the test project that is intended to test the Test.h include file. Let's reorganize the project structure so that both main.cpp and the new Creation.cpp files are in a folder called tests. These will form the basis for a testing executable that exercises all the tests needed to test Test.h. In other words, we're turning the console project that we have into a test project designed to test the test library. The test library is not a separate project because it's just a single header file that will be included as part of other projects.

Later on, in other projects of your own, you can do the same thing. You'll have your primary project with all its source files. You'll also have another test project in a subfolder called tests with its own main.cpp and all the test files. Your test project will include Test.h from the test library but it won't be trying to test the test library as we're doing here. It will instead be focused on testing your own project in the primary project folder. You'll see how all this works once we get the test library to a suitable state so that it can be used to create a different project. We'll be creating a logging library in *Part 2*, *Logging Library*. The logging library will have a subfolder called tests, as I just described.

Turning back to what we have now, let's reorganize the overall project structure for the test library. You can create the tests folder and move main.cpp into it. Make sure to place the new Creation.cpp file into the tests folder. The project structure should look like this:

```
MereTDD project root folder
    Test.h
    tests folder
        main.cpp
        Creation.cpp
```

The main.cpp file can be simplified like this by removing the test and leaving only main:

```
#include "../Test.h"

int main ()
{
    MereTDD::runTests();

    return 0;
}
```

Now, the new Creation.cpp file only contains the single test we have so far, like so:

```
#include "../Test.h"

#include <iostream>

TEST
{
    std::cout << mName << std::endl;
}
```

However, building the project like so now gives a linker error, because we are including Test.h in both the main.cpp and the Creation.cpp compilation units. As a result, we have two methods that result in duplicate symbols. In order to remove the duplicate symbols, we need to declare both getTests and runTests to be inline, like this:

```
inline std::vector<TestInterface *> & getTests ()
{
    static std::vector<TestInterface *> tests;

    return tests;
}

inline void runTests ()
{
    for (auto * test: getTests())
    {
        test->run();
```

```
        }
    }
```

Now, everything builds and runs again and we get the same result as before. The output displays the name of the single test we have so far:

```
testCanBeCreated
Program ended with exit code: 0
```

The output remains unchanged from before. We haven't added any more tests or changed what the current test does. We have changed how the tests are registered and run, and we have reorganized the project structure.

Summary

This chapter has introduced the test library, which consists of a single header file called `Test.h`. It has also shown us how to create a test project, which is a console application that will be used to test the test library.

We have seen how this has evolved from a simple function into a test library that knows how to register and run tests. It's not ready yet. We still have a way to go before the test library can be used in a TDD process to help you design and test your own projects.

By seeing how the test library evolves, you'll come to understand how to use it in your own projects. In the next chapter, you'll understand the challenges of adding multiple tests. There's a reason why we only have a single test so far. Enabling multiple tests and reporting the results of the tests is what the next chapter will cover.

2
Test Results

So far, we have a test library that can only have a single test. You'll see what happens in this chapter when we try to add another test and you'll see how to enhance the test library to support multiple tests. We'll need to use an old and rarely used capability of C++ that actually comes from its early C roots to support multiple tests.

Once we get more than one test, we'll need a way to view the results. This will let you tell at a glance whether everything passed or not. And finally, we'll fix the result output so that it no longer assumes `std::cout`.

We'll cover the following main topics in this chapter:

- Reporting a single test result based on exceptions
- Enhancing the test library to support multiple tests
- Summarizing the test results to clearly see what failed and what passed
- Redirecting the test result so the output can go to any stream

Technical requirements

All code in this chapter uses standard C++ that builds on any modern C++ 17 or later compiler and standard library. The code is based on and continues from the previous chapter.

You can find all the code for this chapter at the following GitHub repository: `https://github.com/PacktPublishing/Test-Driven-Development-with-CPP`.

Reporting a single test result

So far, our single test just prints its hardcoded name when run. There was some early thinking that we might need a result in addition to the test name. This is actually a good example of adding something to the code that is not needed or used. Okay, a minor example because we will need something to keep track of whether the test passes or fails, but it's still a good example of getting ahead of ourselves because we have actually never used the mResult data member yet. We're going to fix that now with a better way to track the result of running a test.

We'll assume that a test succeeds unless something happens to cause it to fail. What can happen? There will eventually be a lot of ways you can cause a test to fail. For now, we'll just consider exceptions. This could be an exception that a test throws on purpose when it detects something is wrong or it could be an unexpected exception that gets thrown.

We don't want any exceptions to stop the tests from running. An exception thrown from one test shouldn't be a reason to stop running others. We still only have a single test but we can make sure that an exception doesn't stop the entire test process.

What we want is to wrap the run function call in a try block so that any exceptions will be treated as a failure, like this:

```
inline void runTests ()
{
    for (auto * test: getTests())
    {
        try
        {
            test->run();
        }
        catch (...)
        {
            test->setFailed("Unexpected exception thrown.");
        }
    }
}
```

When an exception is caught, we want to do two things. The first is to mark the test as a failure. The second is to set a message so that the result can be reported. The problem is that we don't have a method called setFailed on the TestInterface class. It's actually good to first write the code as we'd like it to be.

In fact, the idea of TestInterface was for it to be a set of pure virtual methods like an interface. We could add a new method called setFailed but then the implementation would need to be written in a derived class. This seems like a basic part of a test to be able to hold the result and a message.

So, let's refactor the design and change TestInterface into more of a base class and call it TestBase instead. We can also move the data members from the class declared inside the TEST macro and put them in the TestBase class:

```cpp
class TestBase
{
public:
    TestBase (std::string_view name)
    : mName (name), mPassed (true)
    { }

    virtual ~TestBase () = default;

    virtual void run () = 0;

    std::string_view name () const
    {
        return mName;
    }

    bool passed () const
    {
        return mPassed;
    }

    std::string_view reason () const
    {
        return mReason;
    }

    void setFailed (std::string_view reason)
    {
        mPassed = false;
        mReason = reason;
```

```
        }

    private:
        std::string mName;
        bool mPassed;
        std::string mReason;
    };
```

With the new setFailed method, it no longer made sense to have an mResult data member. Instead, there's an mPassed member, along with the mName member; both came from the TEST macro. It also seemed like a good idea to add some getter methods, especially now that there's also an mReason data member. Altogether, each test can now store its name, remember whether it passed or not, and the reason for failure, if it failed.

Only a slight change is needed in the getTests function to refer to the TestBase class:

```
    inline std::vector<TestBase *> & getTests ()
    {
        static std::vector<TestBase *> tests;

        return tests;
    }
```

The rest of the changes simplify the TEST macro like this to remove the data members, which are now in the base class, and to inherit from TestBase:

```
    #define TEST \
    class Test : public MereTDD::TestBase \
    { \
    public: \
        Test (std::string_view name) \
        : TestBase(name) \
        { \
            MereTDD::getTests().push_back(this); \
        } \
        void run () override; \
    }; \
    Test test("testCanBeCreated"); \
    void Test::run ()
```

Checking to make sure everything builds and runs again shows that we are back to a running program with the same result as before. You'll see this technique often with a refactor. It's good to keep any functional changes to a minimum when refactoring and focus mostly on just getting back to the same behavior as before.

Now, we can make some changes that *will* affect observable behavior. We want to report what is happening while the test is running. For now, we'll just send the output to std::cout. We'll change this later in this chapter to avoid assuming the output destination. The first change is to include iostream in Test.h:

```
#define MERETDD_TEST_H

#include <iostream>
#include <string_view>
#include <vector>
```

Then, change the runTests function to report the progress of the test being run, like this:

```
inline void runTests ()
{
    for (auto * test: getTests ())
    {
        std::cout << "---------------\n"
            << test->name ()
            << std::endl;

        try
        {
            test->run ();
        }
        catch (...)
        {
            test->setFailed ("Unexpected exception thrown.");
        }

        if (test->passed ())
        {
            std::cout << "Passed"
                << std::endl;
```

```
        }
        else
        {
            std::cout << "Failed\n"
                << test->reason()
                << std::endl;
        }
    }
}
```

The original `try`/`catch` remains unchanged. All we do is print some dashes for a separator and the name of the test. It's probably a good idea to flush this line to the output right away. In the case that something happens later, at least the name of the test will be recorded. After the test is run, the test is checked to see whether it passed or not, and the appropriate message is displayed.

We'll also change the test in `Creation.cpp` to throw something to make sure we get a failure. We no longer need to include `iostream` because it's usually not a good idea to display anything from the test itself. You can display output from the test if you want to but any output in the test itself tends to mess up the reporting of the test results. When I sometimes need to display output from within a test, it's usually temporary.

Here is the test modified to throw an int:

```
#include "../Test.h"


TEST
{
    throw 1;
}
```

Normally, you would write code that throws something other than a simple `int` value, but at this point, we just want to show what happens when something does get thrown.

Building and running it now shows the expected failure due to an unexpected exception:

```
- - - - - - - - - - - - - - -
testCanBeCreated
Failed
Unexpected exception thrown.
Program ended with exit code: 0
```

We can remove the `throw` statement from the test so that the body is completely empty and the test will now pass:

```
-----------------

testCanBeCreated
Passed
Program ended with exit code: 0
```

We don't want to keep modifying the test for different scenarios. It's time to add support for multiple tests.

Enhancing the test declaration to support multiple tests

While a single test works, trying to add another one does not build. This is what I tried to do in `Creation.cpp` by adding another test. One of the tests is empty and the second test throws an int. These are the two scenarios we were just trying to work with:

```cpp
#include "../Test.h"


TEST
{
}


TEST
{
    throw 1;
}
```

The failure is due to the `Test` class being declared twice, as well as the `run` method. The TEST macro declares a new global instance of the `Test` class each time it's used. Each instance is called `test`. We don't see the classes or the instances in the code because they are hidden by the TEST macro.

We'll need to modify the TEST macro so that it will generate unique class and instance names. And while we're doing that, let's also fix the name of the test itself. We don't want all tests to have the name `"testCanBeCreated"`, and since the name will need to come from the test declaration, we'll need to also modify the TEST macro to accept a string. Here is how the new `Creation.cpp` file should look:

```cpp
#include "../Test.h"


TEST("Test can be created")
{
```

```
    }

    TEST("Test with throw can be created")
    {
        throw 1;
    }
```

This lets us give sentence names to each test, instead of treating the name like a single-word function name. We still need to modify the TEST macro but it's good to start with the intended usage first and then make it work.

For making unique class and instance names, we could just ask for something unique from the programmer, but the type name of the class and the instance name of that class really are details that the programmer writing tests shouldn't need to worry about. Requiring a unique name to be supplied would only make the details visible. We could instead use a base name and add to it the line number where the test is declared to make both the class and instance names unique.

Macros have the ability to get the line number of the source code file where the macro is used. All we have to do is modify the resulting class and instance names by appending this line number.

It would be nice if this was easy.

All the macros are handled by the preprocessor. It's actually a bit more complicated than that but thinking in terms of the preprocessor is a good simplification. The preprocessor knows how to do simple text replacement and manipulation. The compiler never sees the original code that is written with the macro. The compiler instead sees the end result after the preprocessor is done.

We will need two sets of macros declared in Test.h. One set will generate a unique class name, such as Test7 if the TEST macro was used on line 7. The other set of macros will generate a unique instance name, such as test7.

We need a set of macros because going from a line number to a concatenated result such as Test7 requires multiple steps. If this is the first time you've seen macros used like this, it's normal to find them confusing. Macros use simple text replacement rules that can seem like extra work for us at first. Going from a line number to a unique name requires multiple steps of text replacement that are not obvious. The macros look like this:

```
#define MERETDD_CLASS_FINAL( line ) Test ## line
#define MERETDD_CLASS_RELAY( line ) MERETDD_CLASS_FINAL( line )
#define MERETDD_CLASS MERETDD_CLASS_RELAY( __LINE__ )

#define MERETDD_INSTANCE_FINAL( line ) test ## line
#define MERETDD_INSTANCE_RELAY( line ) MERETDD_INSTANCE_FINAL(
```

```
line )
#define MERETDD_INSTANCE MERETDD_INSTANCE_RELAY( __LINE__ )
```

Each set needs three macros. The macro to use is the last in each set, MERETDD_CLASS and MERETDD_INSTANCE. Each of these will be replaced with the relay macro using the __LINE__ value. The relay macro will see the real line number instead of __LINE__ and the relay macro will then be replaced with the final macro and the line number it was given. The final macro will use the ## operator to do the concatenation. I did warn you that it would be nice if this was easy. I'm sure this is one of the reasons so many programmers avoid macros. At least you've already made it through the most difficult usage of macros in this book.

The end result will be, for example, Test7 for the class name and test7 for the instance name. The only real difference between these two sets of macros is that the class name uses a capital *T* for Test and the instance name uses a lowercase *t* for test.

The class and instance macros need to be added to Test.h right above the definition of the TEST macro that will need to use them. All of this works because, even though the TEST macro looks like it uses many source code lines, remember that each line is terminated with a backslash. This causes everything to end up on a single line of code. This way, all the line numbers will be the same each time the TEST macro is used and the line number will be different the next time it's used.

The new TEST macro looks like this:

```
#define TEST( testName ) \
class MERETDD_CLASS : public MereTDD::TestBase \
{ \
public: \
    MERETDD_CLASS (std::string_view name) \
    : TestBase(name) \
    { \
        MereTDD::getTests().push_back(this); \
    } \
    void run () override; \
}; \
MERETDD_CLASS MERETDD_INSTANCE(testName); \
void MERETDD_CLASS::run ()
```

The MERETDD_CLASS macro is used to declare the class name, declare the constructor, declare the type of the global instance, and scope the run method declaration to the class. All four of these macros will use the same line number because of the backslashes at the end of each line.

The MERETDD_INSTANCE macro is used just once to declare the name of the global instance. It will also use the same line number as the class name.

Building the project and running now shows that the first test passes because it doesn't really do anything and the second test fails because it throws the following:

```
--------------
Test can be created
Passed
--------------
Test with throw can be created
Failed
Unexpected exception thrown.
Program ended with exit code: 0
```

The output ends a bit abruptly and it's time to fix that. We'll add a summary next.

Summarizing the results

The summary can begin with a count of how many tests will be run. I thought about adding a running count for each test but decided against that because the tests are run in no particular order right now. I don't mean that they will be run in a different order each time the testing application is run but they could be reordered if the code is changed and the project is rebuilt. This is because there is no fixed order when creating the final application that the linker will use between multiple .cpp compilation units. Of course, we would need tests spread across multiple files to see the reordering, and right now, all the tests are in Creation.cpp.

The point is that the tests register themselves based on how the global instances get initialized. Within a single .cpp source file, there is a defined order, but there is no guaranteed order between multiple files. Because of this, I decided not to include a number next to each test result.

We'll keep track of how many tests passed and how many failed, and at the end of the for loop that runs all the tests, a summary can be displayed.

As an additional benefit, we can also change the runTests function to return the count of how many tests failed. This will let the main function return the failed count too so that a script can test this value to see whether the tests passed or how many failed. An application exit code of zero will mean that nothing failed. Anything other than zero will represent a failed run and will indicate how many tests have failed.

Here is the simple change to `main.cpp` to return the failed count:

```
int main ()
{
    return MereTDD::runTests();
}
```

Then, here is the new `runTests` function with the summary changes. The changes are described in three parts. All of this is a single function. Only the description is broken into three parts. The first part just displays the count of how many tests will be run:

```
inline int runTests ()
{
    std::cout << "Running "
        << getTests().size()
        << " tests\n";
```

In the second part, we need to keep track of how many tests pass and how many fail, like this:

```
    int numPassed = 0;
    int numFailed = 0;
    for (auto * test: getTests())
    {
        std::cout << "---------------\n"
            << test->name()
            << std::endl;

        try
        {
            test->run();
        }
        catch (...)
        {
            test->setFailed("Unexpected exception thrown.");
        }

        if (test->passed())
        {
            ++numPassed;
```

```
          std::cout << "Passed"
              << std::endl;
      }
      else
      {
          ++numFailed;
          std::cout << "Failed\n"
              << test->reason()
              << std::endl;
      }
  }
```

And in the third part, after looping through all the tests and counting how many passed and how many failed, we display a summary with the counts, like this:

```
  std::cout << "---------------\n";
  if (numFailed == 0)
  {
      std::cout << "All tests passed."
          << std::endl;
  }
  else
  {
      std::cout << "Tests passed: " << numPassed
          << "\nTests failed: " << numFailed
          << std::endl;
  }

  return numFailed;
}
```

Running the project now shows an initial count, the individual test results, and a final summary, and you can also see the application exit code is 1 because of the failed test:

```
Running 2 tests
---------------
Test can be created
Passed
```

```
---------------
Test with throw can be created
Failed
Unexpected exception thrown.
---------------
Tests passed: 1
Tests failed: 1
Program ended with exit code: 1
```

The final line that displays the exit code is not actually part of the testing application. This is normally not shown when the application is run. It's part of the development environment that I am using to write this code. You would normally be interested in the exit code if you were running the testing application from a script such as Python as part of an automated build script.

We have one bit of cleanup still to do with the results. You see, right now, everything gets sent to `std::cout` and this assumption should be fixed so that the results can be sent to any output stream. The next section will do this cleanup.

Redirecting the output results

This is a simple edit that should not cause any real change to the application so far. Right now, the `runTests` function uses `std::cout` directly when displaying the results. We're going to change this so that the `main` function will pass `std::cout` as an argument to `runTests`. Nothing will actually change because we'll still be using `std::cout` for the results but this is a better design because it lets the testing application decide where to send the results, instead of the testing library.

By the testing library, I mean the `Test.h` file. This is the file that other applications will include in order to create and run tests. With the project we have so far, it's a bit different because we're writing tests to test the library itself. So, the whole application is just the `Test.h` file and the `tests` folder containing the testing application.

We first need to change `main.cpp` to include `iostream` and then pass `std::cout` to `runTests`, like this:

```
#include "../Test.h"

#include <iostream>

int main ()
{
```

```
        return MereTDD::runTests(std::cout);
}
```

Then, we no longer need to include iostream in Test.h, because it really doesn't need any input and it doesn't need to refer to std::cout directly. All it needs is to include ostream for the output stream. This could be the standard output, a file, or some other stream:

```
#ifndef MERETDD_TEST_H
#define MERETDD_TEST_H

#include <ostream>
#include <string_view>
#include <vector>
```

Most of the changes are to replace std::cout with a new parameter called output, like this in the runTests function:

```
inline int runTests (std::ostream & output)
{
    output << "Running "
        << getTests().size()
        << " tests\n";

    int numPassed = 0;
    int numFailed = 0;
    for (auto * test: getTests())
    {
        output << "---------------\n"
            << test->name()
            << std::endl;
```

Not all of the changes are shown in the previous code. All you need to do is replace every use of std::cout with output.

This was a simple change and does not affect the output of the application at all. In fact, it's good to make changes like this that are isolated from other changes, just so the new results can be compared with previous results to make sure nothing unexpected has changed.

Summary

This chapter introduced macros and their ability to generate code based on the line number as a way to enable multiple tests. Each test is its own class with its own uniquely named global object instance.

Once multiple tests were supported, then you saw how to track and report the results of each test.

The next chapter will use the build failures in this chapter to show you the first step in the TDD process. We've been following these process steps already without specifically mentioning them. You'll learn more about the TDD process in the next chapter, and the way that the test library has been developed so far should start making more sense as you understand the reasons.

3

The TDD Process

The first two chapters have introduced you to the TDD process by showing you the steps involved. You have seen build failures when declaring multiple tests. You have seen what can happen when we get ahead of ourselves and write code that isn't needed yet. That was a small example with a test result, but it still showed how easy it is to sometimes let code slip into a project before there are tests to support the code. And you also saw how the code starts out with a simple or partial implementation, gets working first, and then is enhanced.

We will cover the following topics in this chapter:

- How build failures come first and should be seen as part of the process
- Why you should write only enough code to pass your tests
- How to enhance a test and get another pass

This chapter will begin by introducing you to the TDD process. For a more detailed walkthrough with more code, refer to *Chapter 10, The TDD Process in Depth*.

Now, it's time to begin learning about the TDD process in a more deliberate manner.

Technical requirements

All code in this chapter uses standard C++ that builds on any modern C++ 17 or later compiler and standard library. The code is based on and continues from the previous chapter.

You can find all the code for this chapter at the following GitHub repository:

`https://github.com/PacktPublishing/Test-Driven-Development-with-CPP`

Build failures come first

In the previous chapter, you saw how the first step to getting multiple tests to run was to write multiple tests. This caused a build failure. When you're programming, it's common to write code that doesn't build at first. These are normally considered mistakes or errors that need to be fixed right away. And gradually, most developers learn to anticipate build errors and avoid them.

When following TDD, I want to encourage you to stop avoiding build errors, because the way to avoid build errors usually means that you work on enabling a new feature or making changes to code before you try to use the new feature or updated code. This means that you're making changes while focused on the details and it is easy to overlook bigger issues, such as how easy it will be to use the new feature or updated code.

Instead, start out by writing code in the way that you think it should be used. That's what was done with the tests. I showed you in the previous chapter that the end result of adding another test should look like this:

```
#include "../Test.h"


TEST
{
}


TEST
{
    throw 1;
}
```

The project was built and was not completed due to errors. This lets us know what needed to be fixed. But before making the changes, I showed what we really wanted the tests to look like:

```
#include "../Test.h"


TEST("Test can be created")
{
}


TEST("Test with throw can be created")
{
    throw 1;
}
```

Changes were made to enable multiple tests only once we had a clear idea of what the tests should look like. Had I not taken this approach, it's possible that some other solution would have been found to name the tests. It might have even worked. But would it have been as easy to use? Would we be able to simply declare a second TEST like the code showed and give each a name right away? I don't know.

But I do know that there have been many times when I did not follow this advice and ended up with a solution that I did not like. I had to go back and redo the work until I was happy with the result. Had I started with the result that I wanted in the first place, then I would have made sure to write code that directly led to that result.

All of this is really just a shift in focus. Instead of diving into the detailed design of what you are coding, take a step back and first write test code to use what you intend to make.

In other words, *let the tests drive the design*. This is the essence of TDD.

The code you write will not build yet because it relies on other code that doesn't exist, but that's OK because it gives you a direction that you'll be happy about.

In a way, writing code from this user point of view gives you a goal and makes that goal real before you even start. Don't settle for a vague idea of what you want. Take the time to write the code as you want it to be used first, build the project, and then work to fix the build errors.

Is it really necessary to try building your project when you know it will fail to build? This is a shortcut that I'll take sometimes, especially when the build takes a long time or when the build failure is obvious. I mean, if I call a method that doesn't exist yet, I'll often write the method next without building. I know it will fail to build and what needs to be done to fix it.

But there are times when this shortcut can lead to problems, such as when working with overloaded methods or template methods. You might write code to use a new overload that doesn't yet exist and think that the code will fail to build, when what actually happens is that the compiler will choose one of the existing overloaded versions to make the call. This is also the case with templates.

You can find a good example of an expected build failure that actually built with no warnings or errors in *Chapter 5, Adding More Confirm Types*. The result was not what was wanted and building first allowed the problem to be caught right away.

The point is that building your project will let you know about these situations. If you expect the build to fail but it compiles anyway, then you know that the compiler was able to figure out a way to make the code work that maybe you weren't expecting. This can lead to valuable insight. Because when you add the intended new overload, it's possible that existing code will start calling your new method too. It's always better to be aware of this situation, rather than being surprised by a hard-to-find bug.

When you're still working on getting your tests to build, you don't need to worry about passing. In fact, it's easier if you let the tests fail at first. Focus on the intended usage instead of getting passing tests.

Once your code builds, how much should you implement? That's the topic of the next section. The main idea is to do as little as possible.

Do only what is needed to pass

When writing code, it's easy to think of all the possibilities of how a method might be used, for example, and to write code to handle each possibility right away. This gets easier with experience and is normally viewed as a good way to write robust code without forgetting to handle different use cases or error conditions.

I urge you to scale back your eagerness to write all this at once. Instead, do only what is needed to pass a test. Then, as you think of other use cases, write a test for each, before extending your code to handle them. The same applies to error cases. As you think of some new error handling that should be added, write a test that will cause that error condition to arise before handling it in your code.

To see how this is done, let's extend the test library to allow for expected exceptions. We have two test cases right now:

```
#include "../Test.h"

TEST("Test can be created")
{
}

TEST("Test with throw can be created")
{
    throw 1;
}
```

The first makes sure that a test can be created. It does nothing and passes. The second test throws an exception. It actually just throws a simple int value of 1. This causes the test to fail. It might seem demotivating to see one or more of your tests fail. But remember, we just got the test to build and that is an accomplishment you should feel good about.

When we initially added the second test in the previous chapter, the goal was to make sure that multiple tests could be added. And the int was thrown to make sure that any exceptions would be treated as a failure. We weren't yet ready to fully handle thrown exceptions. That's what we're going to do now.

We're going to take the existing code that throws an exception and turns it into an expected exception, but we are going to follow the advice given here and do the absolute minimum to get this working. That means we're not going to jump right into a solution that tries to throw multiple different exceptions, and we're not yet going to handle the case where we think an exception should be thrown but it doesn't get thrown.

Because we're writing the testing library itself, our focus sometimes will be on the tests themselves. In many ways, the tests become similar to whatever project-specific code you'll be working with. So, while right now we need to be careful not to add a bunch of tests all at once, you'll want to be careful later not to add a bunch of extra code that doesn't yet have tests all at once. You'll see this shift once we get the test library to a more feature-complete version and then start using it to create a logging library. At that point, the guidance will apply to the logging library and we'll want to avoid adding extra logic to handle different logging scenarios without first adding tests to exercise those scenarios.

Starting with the end usage in mind, we need to think about what the TEST macro usage should look like when there is an expected exception. The main thing we need to communicate is the type of exception that we expect to be thrown.

There will only be one type of exception needed. Even if some code under test throws multiple exception types, we don't want to list more than one exception type per test. That's because, while it's okay for code to check different error conditions and throw a different exception type for each error, each test itself should be written to only test one of these error conditions.

If you have a method that can sometimes throw different exceptions, then you should have a test for each condition that leads to each exception. Each test should be specific and always either lead to a single exception or no exception at all. And if a test expects an exception to be thrown, then that exception should always be thrown in order for the test to be considered to pass.

Later in this chapter, we'll get to the more complicated situation of not catching an exception when one is expected. For now, we want to do only what is needed. Here is what the new usage looks like:

```
TEST_EX("Test with throw can be created", int)
{
    throw 1;
}
```

The first thing you'll notice is that we need a new macro in order to pass the type of exception that is expected to be thrown. I'm calling it TEST_EX, which stands for test exception. Right after the name of the test is a new macro argument for the type of exception that is expected. In this case, it's an int because the code throws 1.

Why do we need a new macro?

Because macros are not really functions. They just work with simple text replacement. We want to be able to tell the difference between a test that doesn't expect any exceptions to be thrown versus a test that does expect an exception to be thrown. Macros don't have the ability to be overloaded like a method or function, with each different version declared with different parameters. A macro needs to be written with a specific number of parameters in mind.

When a test doesn't expect any exception to be thrown, it doesn't make any sense to pass some placeholder value for the exception type. It's better to have one macro that takes just the name and means that no exception is expected, and another macro that takes a name and an exception type.

This is a real example of where the design needs to compromise. Ideally, there would not be a need for a new macro. We're doing the best with what the language gives us here. Macros are an old technology with their own rules.

Going back to the TDD process, you can see that we're again starting with the end usage in mind. Is this solution acceptable? It doesn't exist yet. But if it did, would it feel natural? I think so.

There's no real point in trying to build right now. This is a time when we'll take a shortcut and skip the actual build. In fact, in my editor, the `int` type is already highlighted as an error.

It complains that we're using a keyword wrongly and it might look strange to you as well. You can't just pass types, whether they are keywords or not, as method arguments. Remember that a macro is not really a method though. Once the macro has been fully expanded, the compiler will never see this strange usage of `int`. You can pass types as template parameters. But macros don't support template parameters either.

Now that we have the intended usage, the next step is to think about the solution that will enable this usage. We don't want the test author to have to write a `try`/`catch` block for the expected exception. That's what the test library should do. This means we'll need a new method inside the `Test` class that does have a `try`/`catch` block. This method can catch the expected exception and ignore it for now. We ignore it because we are expecting the exception, which means if we catch it, then the test should pass. If we let the expected exception continue outside of the test, then the `runTests` function will catch it and report a failure due to an unexpected exception.

We want to keep the catch all inside `runTests` because that's how we detect unexpected exceptions. For unexpected exceptions, we don't know what type to catch because we want to be ready to catch anything.

Here, we do know what type of exception to expect because it is being provided in the `TEST_EX` macro. We can have the new method in the `Test` class catch the expected exception. Let's call this new method `runEx`. All the `runEx` method needs to do is look for the expected exception and ignore it. If the test throws something else, then `runEx` won't catch it. But the `runTests` function will be sure to catch it.

Let's look at some code to understand better. Here is the TEST_EX macro in Test.h:

```
#define TEST_EX( testName, exceptionType ) \
class MERETDD_CLASS : public MereTDD::TestBase \
{ \
public: \
    MERETDD_CLASS (std::string_view name) \
    : TestBase (name) \
    { \
        MereTDD::getTests().push_back(this); \
    } \
    void runEx () override \
    { \
        try \
        { \
            run (); \
        } \
        catch (exceptionType const &) \
        { \
        } \
    } \
    void run () override; \
}; \
MERETDD_CLASS MERETDD_INSTANCE (testName); \
void MERETDD_CLASS::run ()
```

You can see that all runEx does is call the original run method inside of a try/catch block that catches the exceptionType specified. In our specific case, we will catch an int and ignore it. All this does is wrap up the run method with a try/catch block so that the test author doesn't have to.

The runEx method is also a *virtual override*. That's because the runTests function needs to call runEx instead of calling run directly. Only then will expected exceptions be caught. We don't want runTests to sometimes call runEx for tests with an expected exception and to call run for those tests without an expected exception. It will be better if runTests always calls runEx.

This means we need to have a default implementation of runEx that just calls run without a try/ catch block. We can do that in the TestBase class, which will need to declare the virtual runEx method anyway. The run and runEx methods look like this inside TestBase:

```
virtual void runEx ()
{
    run();
}


virtual void run () = 0;
```

The TEST_EX macro that expects an exception will override runEx to catch the exception, and the TEST macro that does not expect an exception will use the base runEx class implementation, which just calls run directly.

Now, we need to modify the runTests function to call runEx instead of run, like this:

```
inline int runTests (std::ostream & output)
{
    output << "Running "
        << getTests().size()
        << " tests\n";

    int numPassed = 0;
    int numFailed = 0;
    for (auto * test: getTests())
    {
        output << "---------------\n"
            << test->name()
            << std::endl;

        try
        {
            test->runEx();
        }
        catch (...)
        {
            test->setFailed("Unexpected exception thrown.");
        }
```

Only the first half of the `runTests` function is shown here. The rest of the function remains unchanged. It's really just the single line of code in the `try` block that now calls `runEx` that needed to be updated.

We can now build the project and run it to see how the tests perform. The output looks like this:

```
Running 2 tests
- - - - - - - - - - - - - - -
Test can be created
Passed
- - - - - - - - - - - - - - -
Test with throw can be created
Passed
- - - - - - - - - - - - - - -
All tests passed.
Program ended with exit code: 0
```

The second test used to fail but now it passes because the exception is expected. We also followed the guidance for this section, which is to do only what is needed to pass. The next step in the TDD process is to enhance a test and get another pass.

Enhancing a test and getting another pass

What happens if a test that expects an exception does not see the exception? That should be a failure, and we'll handle it next. This situation is a little different because the next *pass* is really going to be a *failure*.

When you're writing tests and following the guidance to first do the minimum amount to get a first passing result and then enhancing the test to get another pass, you'll be focused on passing. That's good because we want all the tests to eventually pass.

Any failure should almost always be a failure. It doesn't usually make sense to have *expected failures* in your tests. What we're about to do here is a bit out of the ordinary and it's because we're still developing the test library itself. We need to make sure that a missing exception that was expected and did not occur is able to be caught as a failed test. We then want to treat that failed test as a pass because we're testing the ability of the test library to be able to catch these failures.

Right now, we have a hole in the test library because adding a third test that expects an int to be thrown but never actually throws an int is seen as a passing test. In other words, the tests in this set all pass:

```
#include "../Test.h"

TEST("Test can be created")
{
}

TEST_EX("Test with throw can be created", int)
{
    throw 1;
}

TEST_EX("Test that never throws can be created", int)
{
}
```

Building this works okay and running it shows that all three tests pass:

```
Running 3 tests
---------------
Test can be created
Passed
---------------
Test with throw can be created
Passed
---------------
Test that never throws can be created
Passed
---------------
All tests passed.
Program ended with exit code: 0
```

This is not what we want. The third test should fail because it expected an int to be thrown but that did not happen. But that also goes against the goal that all tests should pass. There is no way to have an expected failure. Sure, we might be able to add this concept into the testing library, but that would add extra complexity.

If we were to add the ability for a test to fail but still be treated as if it passed, then what would happen if the test failed for some unexpected reason? It would be easy for a bad test to be written that fails for multiple reasons but actually gets reported as a pass because the failure was expected.

While writing this, I initially decided not to add the ability to have expected failures. My reasoning was that all tests should pass. But that left us in a bind, because how else can we verify that the test library itself can properly detect missing expected exceptions?

We need to close the hole exposed by the third test.

There is no good answer to this dilemma. So, what I'm going to do is get this new test to fail and then add the ability to treat a failure as a success. I don't like the alternatives, which is to leave the test in the code but comment it out so that it wouldn't actually run, or to delete the third test entirely.

What finally convinced me to add support for successful failing tests was the idea that everything should be tested, especially big things, such as the ability to make sure that an expected exception is always thrown. You probably won't need to use the ability to mark a test as an expected failure but if you do, then you will be able to do the same thing. We are in a unique situation because we need to test something about the test library itself.

Alright, let's get the new test to fail. The minimum amount of code needed for this is to return if the expected exception was caught. If the exception was not caught, then we throw something else. The code to update is the TEST_EX macro override of the runEx method, like this:

```
void runEx () override \
{ \
    try \
    { \
        run(); \
    } \
    catch (exceptionType const &) \
    { \
        return; \
    } \
    throw 1; \
} \
```

The rest of the macro is unchanged, so only the `runEx` override is shown here. We return when the expected exception is caught, which will cause the test to pass. And after the `try/catch` block, we throw something else that will cause the test to fail.

If you find it strange to see a simple int value being thrown, remember that our goal is to do the absolute minimum needed at this point. You would never want to leave code that throws something like this and we will fix that next.

This works and is great because it is the minimum amount needed to do what we want, but the result looks strange and misleading. Here is the test result output:

```
Running 3 tests
---------------
Test can be created
Passed
---------------
Test with throw can be created
Passed
---------------
Test that never throws can be created
Failed
Unexpected exception thrown.
---------------
Tests passed: 2
Tests failed: 1
Program ended with exit code: 1
```

You can see that we got a failure but the message says `Unexpected exception thrown.`. This message is almost the exact opposite of what we want. We want it to say that an expected exception was not thrown. Let's fix this before we continue turning it into an expected failure.

First, we need some way for the `runTests` function to detect the difference between an unexpected exception and a missing exception. Right now, it just catches everything and treats any exception as unexpected. If we were to throw something special and catch it first, then that could be the signal that an exception was missing. And anything else that gets caught would be unexpected. OK, what should this special throw be?

The best thing to throw is going to be something that the test library defines specifically for this purpose. We can define a new class just for this.

Let's call it `MissingException` and define it inside the `MereTDD` namespace, like this:

```
class MissingException
{
public:
    MissingException (std::string_view exType)
    : mExType (exType)
    { }

    std::string_view exType () const
    {
        return mExType;
    }

private:
    std::string mExType;
};
```

Not only will this class signal that an expected exception was not thrown but it will also keep track of the type of exception that should have been thrown. The type will not be a real type in the sense that the C++ compiler understands types. It will be the text representation of that type. This actually fits well with the design because that's what the `TEST_EX` macro accepts, a piece of text that gets substituted in the code for the actual type when the macro is expanded.

Inside the `TEST_EX` macro implementation of the `runEx` method, we can change it to look like this:

```
    void runEx () override \
    { \
        try \
        { \
            run (); \
        } \
        catch (exceptionType const &) \
        { \
            return; \
        } \
        throw MereTDD::MissingException (#exceptionType); \
    } \
```

Instead of throwing an int like before, the code now throws a `MissingException`. Notice how it uses another feature of macros, which is the ability to turn a macro parameter into a string literal with the # operator. By placing # before `exceptionType`, it will turn the `int` provided in the `TEST_EX` macro usage into an `"int"` string literal, which can be used to initialize the `MissingException` with the name of the type of exception that is expected.

We're now throwing a special type that can identify a missing exception, so the only piece remaining is to catch this exception type and handle it. This happens in the `runTests` function, like this:

```
try
{
    test->runEx();
}
catch (MissingException const & ex)
{
    std::string message = "Expected exception type ";
    message += ex.exType();
    message += " was not thrown.";
    test->setFailed(message);
}
catch (...)
{
    test->setFailed("Unexpected exception thrown.");
}
```

The order is important. We need to first try catching `MissingException` before catching everything else. If we do catch `MissingException`, then the code changes the message that gets displayed to let us know what type of exception was expected but not thrown.

Running the project now shows a more applicable message for the failure, like this:

```
Running 3 tests
---------------
Test can be created
Passed
---------------
Test with throw can be created
Passed
---------------
Test that never throws can be created
```

```
Failed
Expected exception type int was not thrown.
---------------
Tests passed: 2
Tests failed: 1
Program ended with exit code: 1
```

This clearly describes why the test failed. We now need to turn the failure into a passing test and it will be nice to keep the failure message. We'll just change the status from **Failed** to **Expected failure**. Since we're keeping the failure message, I have an idea for something that will make this ability to mark failed tests as passing a safer feature.

What do I mean by a safer feature? Well, this was one of my biggest concerns with adding the ability to have expected failures. Once we mark a test as an expected failure, then it would be too easy for the test to fail for other reasons. Those other reasons should be treated as real failures because they were not the expected reason. In other words, if we just treat any failure as if a test passes, then what happens if a test fails for a different reason? That would be treated as a pass too and that would be bad. We want to mark failures as passing but only for expected failures.

In this particular case, if we were to just treat a failure as a pass, then what would happen if the test was supposed to throw an int but instead threw a string? That would definitely cause a failure and we need a test case for this too. We might as well add that test now. We don't want to treat the throwing of a different exception the same as not throwing any exception at all. Both are failures but the tests should be specific. Anything else should cause a legitimate failure.

Let's start with the end usage in mind and explore how best to express the new concept. I thought about adding an expected failure message to the macro but that would require a new macro. And really, it would require a new macro for each macro we already have. We'd need to extend both the TEST macro and the TEST_EX macro with two new macros, such as FAILED_TEST and FAILED_TEST_EX. That doesn't seem like a very good idea. What if, instead, we add a new method to the TestBase class? It should look like this when used in the new tests:

```
// This test should fail because it throws an
// unexpected exception.
TEST("Test that throws unexpectedly can be created")
·{
    setExpectedFailureReason(
        "Unexpected exception thrown.");

    throw "Unexpected";
}
```

```
// This test should fail because it does not throw
// an exception that it is expecting to be thrown.
TEST_EX("Test that never throws can be created", int)
{
    setExpectedFailureReason(
        "Expected exception type int was not thrown.");
}

// This test should fail because it throws an
// exception that does not match the expected type.
TEST_EX("Test that throws wrong type can be created", int)
{
    setExpectedFailureReason(
        "Unexpected exception thrown.");

    throw "Wrong type";
}
```

Software design is all about trade-offs. We are adding the ability to have a failing test turn into a passing test. The cost is extra complexity. Users need to know that the setExpectedFailureReason method needs to be called inside the test body to enable this feature. But the benefit is that we can now test things in a safe manner that would not have been possible otherwise. The other thing to consider is that this ability to set expected failures will most likely not be needed outside of the test library itself.

Expected failure reasons are also a little hard to get right. It's easy to miss something, such as a period at the end of the failure reason. The best way I found to get the exact reason text is to let the test fail and then copy the reason from the summary description.

Until now, we haven't been able to have a test that specifically looks for a completely unexpected exception. Now, we can. And for the times when we expect an exception to be thrown, we can now check the two failure cases that go along with this, when the expected type is not thrown and when something else is thrown.

All of this is better than the alternative of either leaving these tests out or commenting them out, and we can do all this without adding more macros. Of course, the tests won't compile yet because we haven't created the setExpectedFailureReason method. So, let's add that now:

```
std::string_view reason () const
{
    return mReason;
}
```

```cpp
    std::string_view expectedReason () const
    {
        return mExpectedReason;
    }

    void setFailed (std::string_view reason)
    {
        mPassed = false;
        mReason = reason;
    }

    void setExpectedFailureReason (std::string_view reason)
    {
        mExpectedReason = reason;
    }

private:
    std::string mName;
    bool mPassed;
    std::string mReason;
    std::string mExpectedReason;
};
```

We need a new data member to hold the expected reason, which will be an empty string, unless set inside the test body. We need the setExpectedFailureReason method to set the expected failure reason and we also need an expectedReason getter method to retrieve the expected failure reason.

Now that we have this ability to mark tests with a specific failure reason that is expected, let's look for the expected failures in the runTests function:

```cpp
        if (test->passed())
        {
            ++numPassed;
            output << "Passed"
                << std::endl;
        }
        else if (not test->expectedReason().empty() &&
```

```
                        test->expectedReason() == test->reason())
            {

                ++numPassed;
                output << "Expected failure\n"
                    << test->reason()
                    << std::endl;
            }
            else
            {

                ++numFailed;
                output << "Failed\n"
                    << test->reason()
                    << std::endl;
            }
```

You can see the new test for tests that did not pass in the `else if` block. We first make sure that the expected reason is not empty and that it matches the actual failure reason. If the expected failure reason matches the actual failure reason, then we treat this test as a pass because of an expected failure.

Building the project and running it now shows that all five tests are passing:

```
Running 5 tests
---------------
Test can be created
Passed
---------------
Test that throws unexpectedly can be created
Expected failure
Unexpected exception thrown.
---------------
Test with throw can be created
Passed
---------------
Test that never throws can be created
Expected failure
Expected exception type int was not thrown.
---------------
Test that throws wrong type can be created
```

```
Expected failure
Unexpected exception thrown.
- - - - - - - - - - - - - - -
All tests passed.
Program ended with exit code: 0
```

You can see the three new tests that have expected failures. All of these are passing tests and we now have an interesting ability to expect tests to fail. Use it wisely. It is not normal to expect tests to fail.

We still have one more scenario to consider. And I'll be honest and say that I took a break for an hour or so before I thought of this. We need to make sure the test library covers everything we can think of because you'll be using it to test your code. You need to have a high level of confidence that the test library itself is as bug-free as possible.

Here's the case we need to handle. What if there is a test case that's expected to fail for some reason but it actually passes? Right now, the test library first checks whether the test has passed, and if so, then it doesn't even look to see whether it was supposed to have failed. If it passes, then it passes.

But if you go to all the trouble to set an expected failure reason and the test passes instead, what should be the outcome? What we have is a failure that should have been treated as a pass that actually passed instead. Should this be a failure after all? A person could go dizzy thinking about these things.

If we treat this as a failure, then we're back to where we started with a test case that we want to include but that is designed to ultimately fail. And that means we either have to live with a failure in the tests, ignore the scenario and skip the test, write the test and then comment it out so it doesn't normally run, or find another solution.

Living with a failure is not an option. When using TDD, you need to get all your tests to a passing state. It does no good to expect a failure. That's the whole reason we went to all the trouble of allowing failing tests to be expected to fail. Then, we can call those failures passes because they were expected.

Skipping a test is also not an option. If you decide something is really not an issue and doesn't need to be tested, then that's different. You don't want a bunch of useless tests cluttering up your project. This seems like something important that we don't want to skip, though.

Writing a test and then disabling it so it doesn't run is also a bad idea. It's too easy to forget the test ever existed.

We need another solution. And no, it's not going to be to add another level where a passing test that should have failed in an expected way instead is treated as a failure, which we will then somehow mark as passing again. I'm not even sure how to write that sentence, so I'm going to leave it as confusing as it sounds. That path leads to a never-ending cycle of pass-fail-pass-fail-pass thinking. Too complicated.

The best idea I can come up with is to treat this case as a **missed failure**. That will let us test for the scenario and always run the test but avoid the true failure that would cause automated tools to reject the build, due to failures found.

Here is the new test that shows the scenario just described. It will currently pass without any problems:

```
// This test should throw an unexpected exception
// but it doesn't. We need to somehow let the user
// know what happened. This will result in a missed failure.
TEST("Test that should throw unexpectedly can be created")
{
    setExpectedFailureReason(
        "Unexpected exception thrown.");
}
```

Running this new test does indeed pass unnoticed like this:

```
Running 6 tests
- - - - - - - - - - - - - - -
Test can be created
Passed
- - - - - - - - - - - - - - -
Test that throws unexpectedly can be created
Expected failure
Unexpected exception thrown.
- - - - - - - - - - - - - - -
Test that should throw unexpectedly can be created
Passed
- - - - - - - - - - - - - - -
Test with throw can be created
Passed
- - - - - - - - - - - - - - -
Test that never throws can be created
Expected failure
Expected exception type int was not thrown.
- - - - - - - - - - - - - - -
Test that throws wrong type can be created
```

```
Expected failure
Unexpected exception thrown.
----------------
All tests passed.
Program ended with exit code: 0
```

We need to check in the `runTests` function for passed tests whether the expected error result has been set and, if so, then increment a new `numMissedFailed` count instead of the passed count. The new count should be summarized at the end too, but only if it's anything other than zero.

Here is the beginning of `runTests`, where the new `numMissedFailed` count is declared:

```
inline int runTests (std::ostream & output)
{
    output << "Running "
        << getTests().size()
        << " tests\n";

    int numPassed = 0;
    int numMissedFailed = 0;
    int numFailed = 0;
```

Here is the part of `runTests` that checks for passing tests. Inside of here is where we need to look for a passing test that was supposed to have failed with an expected failure reason:

```
        if (test->passed())
        {
            if (not test->expectedReason().empty())
            {
                // This test passed but it was supposed
                // to have failed.
                ++numMissedFailed;
                output << "Missed expected failure\n"
                    << "Test passed but was expected to fail."
                    << std::endl;
            }
            else
            {
                ++numPassed;
```

```
                output << "Passed"
                    << std::endl;
            }
        }
```

And here is the end of the `runTests` function that summarizes the results. This will now show the test failures missed, if there are any:

```
output << "---------------\n";
output << "Tests passed: " << numPassed
        << "\nTests failed: " << numFailed;
if (numMissedFailed != 0)
{
    output << "\nTests failures missed: " <<
    numMissedFailed;
}
output << std::endl;

return numFailed;
}
```

The summary at the end started getting more complicated than it needed to be. So, it now always shows the passed and failed count and only the failures missed if there were any. We now get a missed failure for the new test that was expected to fail but ended up passing.

Should the missed failures be included in the failure count? I thought about this and decided to only return the number of actual failures for all the reasons just explained that led to this scenario in the first place. Remember that it's highly unlikely that you will ever find yourself needing to write a test that you intend to fail and then treat as a pass. So, you should not have missed failures either.

The output looks like this:

```
Running 6 tests
---------------
Test can be created
Passed
---------------
Test that throws unexpectedly can be created
Expected failure
Unexpected exception thrown.
```

```
----------------
Test that should throw unexpectedly can be created
Missed expected failure
Test passed but was expected to fail.
----------------
Test with throw can be created
Passed
----------------
Test that never throws can be created
Expected failure
Expected exception type int was not thrown.
----------------
Test that throws wrong type can be created
Expected failure
Unexpected exception thrown.
----------------
Tests passed: 5
Tests failed: 0
Tests failures missed: 1
Program ended with exit code: 0
```

We should be good now for this part. You have the ability to expect an exception and rely on your test to fail if the exception is not thrown, and the test library fully tests itself with all the possible combinations around expected exceptions.

This section also demonstrated multiple times how to continue enhancing your tests and getting them to pass again. If you follow this process, you'll be able to gradually build your tests to cover more complicated scenarios.

Summary

This chapter has taken the steps we've already been following and made them explicit.

You now know to write code the way you want it to be used first, instead of diving into the details and working from the bottom up in order to avoid build failures. It's better to work from the top, or an end user point of view, so that you will have a solution you'll be happy with, instead of a buildable solution that is hard to use. You do this by writing tests as you would like your code to be used. Once you are happy with how your code will be used, then build it and look at the build errors to fix them. Getting the tests to pass is not the goal yet. This slight change in focus will lead to designs that are easier and more intuitive to use.

Once your code builds, the next step is to do only what is needed to get the tests to pass. It's always possible that a change will cause tests that used to pass to now fail. That's okay and is another good reason to do only what is needed.

And finally, you can enhance a test or add more tests before writing the code needed to pass everything again.

The test library is far from complete though. The only way to cause tests to fail right now is to throw an exception that is not expected. You can see that, even at a higher level, we're following the practice of doing only what is needed, getting that working, and then enhancing the tests to add more.

The next enhancement is to let the test programmer check conditions within a test to make sure everything is working correctly. The next chapter will begin this work.

4

Adding Tests to a Project

In this chapter, we're going to add a major new ability to the test library. The new ability will let you check conditions within a test to make sure everything is going as planned. Sometimes, these checks are called an *assert*, and sometimes, they are called an *expect*. Whatever they are called, they let you confirm that the values you get back from the code being tested match expectations.

For this book and the test library that we're creating, I'm going to call these checks confirmations. Each confirmation will be called a *confirm*. The reason for this is that assert is already being used in C++, and it can be confusing to use the same name. Additionally, expect is a common term within other test libraries, which is not by itself a reason to avoid using the same term. I actually like the term expect. But expect has another common behavior that we don't want. Many other testing libraries will let a test continue even if an expect fails. I don't really like this behavior. Once something has gone wrong, I think it's time to end that test. Other tests can still run. But we shouldn't continue running a test once something doesn't match what we expect.

So far, you can use the test library to write multiple tests, run them, and see the results. The result of each test is to either pass or fail. You can even expect certain failures and treat them as passing. And there's a third result that will likely not be needed outside of the test library itself and that is a missed failure. You can read all about these abilities in the first three chapters.

In this chapter, we will cover the following main topics:

- How to detect whether a test passes or fails

- Enhancing the testing library to support confirmations

- Should error cases be tested, too?

There's a reason we've waited until this chapter to add confirms. We're following a TDD approach to the design of the test library itself. That means we let the tests drive the design. This is an agile approach to software design. We think about what is the most valuable or necessary feature or capability to add next, what the end use of that feature will be, write the minimum amount of code needed to get it working, and then enhance the design by adding more.

Until now, there was no point in adding confirms. We needed to get the essential functionality working first, which would let tests be created and run before we could think about what to do inside the tests. Maybe we could have added confirms before the exception handling. But I choose to work on exception handling before confirms. Exceptions seem more closely related to the essential declaration and running of the tests than confirms and, therefore, are more valuable than confirms.

Additionally, you'll also see that we'll be using exceptions to enable confirmations. This is another reason why the basic ability to handle exceptions came before confirms.

Now we can turn our attention to the tests with confirms. Again, we're going to do the minimum amount of work needed to get the confirms functional and useful. We'll continue adding more abilities to confirms in the next chapter.

Technical requirements

All of the code in this chapter uses standard C++, which builds on any modern C++ 17, or later, compiler and standard library. The code is based on and continues from the previous chapters.

You can find all of the code for this chapter at the following GitHub repository:

https://github.com/PacktPublishing/Test-Driven-Development-with-CPP

How to detect whether a test passes or fails

In this chapter, the tests we'll be creating are different enough from the creation tests that they should have their own file. When writing your own tests, you'll want to organize them into multiple files, too.

Let's create a new file called `Confirm.cpp` and place it inside the `tests` folder. With the new file, the project structure will look like the following:

```
MereTDD project root folder
    Test.h
    tests folder
        main.cpp
        Confirm.cpp
        Creation.cpp
```

Then, add a single test to the new file so that it looks like this:

```
#include "../Test.h"

TEST("Test will pass without any confirms")
{
}
```

We already have an empty test in `Creation.cpp`, which looks like this:

```
TEST("Test can be created")
{
}
```

The only real difference between these two tests is the name. Do we really need another test that does the exact same thing but with a different name? I could argue on either side of a debate about adding code that does the same thing but with a different name. Some people might see this and think it is pure code duplication.

To me, the difference comes down to *intent*. Yes, both tests happen to be the same right now. But who knows whether one or both will be modified later? And if that ever happens, will we be able to remember that a test was serving multiple purposes?

I strongly urge you to write each test as if it is the only thing standing between your code and the bugs that the test is designed to prevent. Or maybe the test is exercising a specific usage to make sure that nothing breaks later during a design change. Having two identical tests is okay as long as they are testing different things. It's the goal that should be unique.

In this case, the original test just ensures that a test can be created in its most basic form. The new test is specifically making sure that an empty test will pass. These are two different tests that just happen to require the same test method body to accomplish their goals.

Now that we have a new file in the project with a new test, let's build and make sure everything works as expected. And it fails. The reason for the build failure is the following:

ld: 5 duplicate symbols for architecture x86_64

Everything compiles okay, but the project fails to link into the final runnable executable. We have five linker errors. One of the linker errors says the following:

duplicate symbol 'Test3::run()'

I have only listed one of the linker errors because they are all similar. The problem is that we now have two declarations of `Test3`. One declaration comes from each file, `Creation.cpp` and `Confirm.cpp`; that is because the `TEST` macro declares the `Test` class with a unique number based on the line number where the `TEST` macro appears in the source file. Both files happen to use the `TEST` macro on line 3, so they each declare a class called `Test3`.

The solution for this is to use an *unnamed namespace* in the macros when declaring the class. This will still create two classes such as `Test3`, but each will be in a namespace that does not extend outside of the `.cpp` file in which it is declared. This means that the test classes can continue to be based on the line number, which is guaranteed to be unique within each `.cpp` file and will now no longer conflict with any other tests that happen to be declared on the same line number in a different `.cpp` file.

All we need to do is modify the TEST and TEST_EX macros to add an unnamed namespace around just the class declaration inside of each macro. We don't need to extend the namespace to the end of the macro because the macros go on to declare the beginning of the run method. Luckily, the run method declaration does not need to be inside the namespace. Otherwise, we would have to figure out how to end the namespace with the closing curly brace after the run method has been fully defined. As it is, we can end the namespace after the class declaration. The TEST macro looks like this:

```
#define TEST( testName ) \
namespace { \
class MERETDD_CLASS : public MereTDD::TestBase \
{ \
public: \
    MERETDD_CLASS (std::string_view name) \
    : TestBase(name) \
    { \
        MereTDD::getTests().push_back(this); \
    } \
    void run () override; \
}; \
} /* end of unnamed namespace */ \
MERETDD_CLASS MERETDD_INSTANCE(testName); \
void MERETDD_CLASS::run ()
```

And the TEST_EX macro needs a similar unnamed namespace as follows:

```
#define TEST_EX( testName, exceptionType ) \
namespace { \
class MERETDD_CLASS : public MereTDD::TestBase \
{ \
public: \
    MERETDD_CLASS (std::string_view name) \
    : TestBase(name) \
    { \
        MereTDD::getTests().push_back(this); \
    } \
    void runEx () override \
    { \
        try \
```

```
    { \
        run(); \
    } \
    catch (exceptionType const &) \
    { \
        return; \
    } \
    throw MereTDD::MissingException(#exceptionType); \
  } \
  void run () override; \
}; \
} /* end of unnamed namespace */ \
MERETDD_CLASS MERETDD_INSTANCE(testName); \
void MERETDD_CLASS::run ()
```

Now that the project builds again, running it will show the new test. Depending on what order your linker built the final executable, you might find the new test runs before or after the previous tests. Here is a portion of the results when I ran the test project:

```
Running 7 tests
---------------
Test will pass without any confirms
Passed
---------------
Test can be created
Passed
---------------
```

The other five tests and the summary are not shown. The previous chapter ended with six tests, and we just added one more bringing the total to seven tests. The important part is that the new test runs and passes. Now we can think about what a confirm will look like. And what does it mean to confirm something?

When running a test, you want to not just verify that the test completes but that it completes correctly. And it also helps to check along the way to make sure everything is running as expected. You can do this by comparing the values you get from the code being tested to make sure they match the expected values.

Let's say that you have a function that adds two numbers and returns a result. You can call this function with known values and compare the returned sum with an expected sum that you calculate yourself. You confirm that the calculated sum matches the expected sum. If the values match, then the confirm passes. But if the values don't match, then the confirm fails, which should cause the test to fail, too.

A test can have multiple confirms and each will be checked to make sure they pass. The moment one confirm fails, there's no point in continuing the test because it has already failed. Some TDD purists will claim that a test should only have a single confirmation. I think there's a good compromise between only having a single confirm versus writing epic tests that try to verify everything.

A popular style of writing tests with multiple confirmations is to keep track of how many confirms pass by letting a test continue even if a confirm fails. There is a benefit to this style because the developer can sometimes fix multiple problems with a single run of the tests. We're not taking this approach because I think the benefit is rarely achieved in practice. Some people might argue this, but hear me out. Once something is proven to not meet your expectations, the most likely result is a chain reaction of further failures. I have rarely seen a well-designed test fail one confirmation and then somehow recover to pass unrelated confirmations. If the test behaves like this, then it normally is testing unrelated issues and should be broken into multiple tests. The practice we're going to be following is this: when a confirm fails, then the test itself has failed. Other tests might proceed just fine. But the test with a failed confirm has already failed, and there is no point in continuing to see whether maybe some part of the test might still be okay.

When writing tests, just like when writing regular code, it's good to avoid duplication. In other words, if you find yourself testing the same things by checking values that have already been checked in other tests, then it's time to think about the goal of each test. Write one test that covers some basic functionality that will be used many times. Then, in other tests that make use of that same functionality, you can assume it has already been tested and works, so there is no need to verify it again with extra confirms.

Some code will probably make all of this clearer. First, let's think about how to verify an expected result without a confirm. This is a time when we can't just write the code for what a confirm will look like because we don't know yet what we want it to do. A little exploration is in order. The next section will turn the exploration we'll do here into actual confirms.

For a moment, let's pretend that we have a real TDD project that we're working on. We'll keep things simple and say that we need some way to determine whether a school grade is passing or not. Even this simple example could become complicated if there were different guidelines for passing homework versus quizzes or tests. If that were the case, there might be a whole class hierarchy involved. We just have a simple need to determine whether a score from 0 to 100 is a passing grade or not.

Now that we have our scenario, what would a simple test look like? We don't have any code to support the grading requirement. It's just a general idea of what we want. So, we expect the build to fail if we try running right after creating a test. This is how you can use TDD to come up with a design.

For now, we'll put this code inside `Confirm.cpp`. If we were really building a test project for a school grading application, then there might be a test file called `Grades.cpp`. Because we're just exploring, we'll use the test file we already have, called `Confirm.cpp`, and create a test like this:

```
TEST("Test passing grades")
{
    bool result = isPassingGrade(0);
    if (result)
    {
        throw 1;
    }
}
```

The first thing is to think about the usage. If you had a function called `isPassingGrade` that accepted a score and returned a bool result, would that meet your requirements and be easy to use? It seems easy enough. It will do whatever it needs inside to tell us whether the score is passing or not and return true if the grade is passing and false if it's not.

Then, you can think about how to test this function. It's always good to test boundary conditions, so we can start by asking whether a score of 0 is passing or not. We assign the passing result to a variable that can be tested against an expected value. We expect 0 to be a failing grade, which is why the code throws something if the result is true instead. This will cause the test case to fail because of an unexpected exception.

We're on the right track. This is what I want you to understand about checking along the way to make sure everything is running okay. We could add another check in the same test like this to make sure that 100 is a passing grade:

```
TEST("Test passing grades")
{
    bool result = isPassingGrade(0);
    if (result)
    {
        throw 1;
    }

    result = isPassingGrade(100);
    if (not result)
    {
        throw 1;
```

```
        }
    }
```

Now, you can see a single test that checks two things. First, it makes sure that a score of 0 is a failing grade and then that a score of 100 is a passing grade. Because these checks are so related, I would put them in the same test as this and confirm that the first case should be a failing grade and the second should be a passing grade.

A test confirmation is nothing more than a simple check against an expected value that throws an exception if the expectation is not met.

Some TDD purists will recommend that you split the test into two separate tests. My advice is to use your best judgment. I tend to avoid absolute guidance that says you should *always* do something one way or another. I think there's room to be flexible.

Let's get this building so that we can run it and see the results. All we need to do is add the isPassingGrade function. We'll add the function to the top of Confirm.cpp. If this was a real project you were working on, then you would have a better place to put this function. It would not be in the test project; instead, it would be included in the project being tested.

Inside Confirm.cpp, create a function called isPassingGrade, as follows:

```
bool isPassingGrade (int value)
{
    return true;
}
```

Now we can build and run the project to see the results. The test result we're interested in fails like this:

```
- - - - - - - - - - - - - - -
Test passing grades
Failed
Unexpected exception thrown.
- - - - - - - - - - - - - - -
```

The function should obviously fail because it always returns true for a passing grade regardless of the score given. But that's not the part we're going to focus on next. It would be if you really were building and testing a grading application. You would enhance the design, get the test to pass, and then enhance the test, and continue until all the tests pass.

This is enough to demonstrate what I mean by checking on the progress of a running test to make sure it's proceeding as expected. Now we have a test that, first, checks to make sure 0 is a failing grade and then checks to make sure 100 is a passing grade. Each of these is a confirm. At each point, we are checking whether the actual result matches what we expect. And we confirm in different ways to fit each condition.

In the next section, we're going to enhance the test library to fix problems with the current solution and make it easier to write confirms. Right now, the code throws an int when it detects a problem, and while the throw definitely causes the test to fail, it leads to a test result that says the failure was caused by an unexpected exception.

The next section will wrap the if statement with its criteria and the exception throwing into an easy macro that will handle everything and lead to a better description of what actually failed and where it failed.

Enhancing the testing library to support assertions

The passing grades test from the previous section has two confirms that we're going to improve in this section. It looks like this:

```
TEST("Test passing grades")
{
    bool result = isPassingGrade(0);
    if (result)
    {
        throw 1;
    }

    result = isPassingGrade(100);
    if (not result)
    {
        throw 1;
    }
}
```

In the first confirm, we want to make sure that result is false because we know that a score of 0 should not be a passing grade. And for the second confirm, we want to make sure that, this time, result is true because we know that a score of 100 should lead to a passing grade.

Can you see how the if condition needs to be the *opposite* of what we're trying to confirm? This is because the if block runs when the confirm does *not* meet the expected value. We'll need to make this easier to use because it will lead to bugs if we always have to write confirms like this. But there are still bigger problems with the test code.

Why does it throw an int if the check fails? That's because we're still exploring what a real confirm should look like. The code we have now just shows you the need for making checks along the way inside of a test to ensure things are proceeding as expected. This section will change how we're going to be writing confirms in our tests.

Throwing an int when a value does not match what was expected also leads to the wrong test result description. We don't want the test results to say that an unexpected exception was thrown.

However, we do want to throw something. Because once a test deviates from the expected path, we don't want the test to continue. It's already shown that it has failed. Throwing whenever an expected condition is not met is a great way to fail the test at that point. We need to figure out a way to change the test result description to better inform us of what went wrong.

First, let's fix the test result by throwing something more meaningful. Note that the following code uses hardcoded numeric values, such as 17 and 23. Numbers such as these are often called *magic numbers* and should be avoided. We'll be fixing the problem soon, and the use of direct numbers whose meaning is unclear is included to show you that there is a better way. In Confirm.cpp, change the passing grades test to throw BoolConfirmException from both confirms like this:

```
TEST("Test passing grades")
{
    bool result = isPassingGrade(0);
    if (result)
    {
        throw MereTDD::BoolConfirmException(false, 17);
    }

    result = isPassingGrade(100);
    if (not result)
    {
        throw MereTDD::BoolConfirmException(true, 23);
    }
}
```

Later, we'll need to create this class. For now, we want to code it like we intend to use it. It's called BoolConfirmException because it will let us confirm that a bool value matches what we expect. The constructor parameters will be the expected bool value and the line number. I used line numbers 17 and 23 because they are the line numbers in my editor for the two throw statements. Later in this section, we'll use a macro so that we can let the macro provide the line number automatically. Normally, you would want to avoid hardcoding any numeric value in the code except for simple values such as 0, 1, and maybe -1. Any other values are called magic numbers because the meaning is confusing.

The exception thrown in confirms will be based on the information needed to make a meaningful test result description. For bool values, the expected value and line number are enough. Other exceptions will need more information and will be explained in the next chapter. We'll have more than one exception type, but they will be related. Inheritance is a good way to represent the different exception types that we'll be throwing. The base class for all the types will be called ConfirmException.

In Test.h, create a new class called ConfirmException inside the MereTDD namespace like this:

```
namespace MereTDD
{

class ConfirmException
{
public:
    ConfirmException () = default;
    virtual ~ConfirmException () = default;

    std::string_view reason () const
    {
        return mReason;
    }

protected:
    std::string mReason;
};
```

Then, right after the base exception class, we can declare the derived BoolConfirmException class like this:

```
class BoolConfirmException : public ConfirmException
{
public:
    BoolConfirmException (bool expected, int line)
    {
        mReason =  "Confirm failed on line ";
        mReason += std::to_string(line) + "\n";
        mReason += "    Expected: ";
        mReason += expected ? "true" : "false";
```

```
        }
    };
```

The purpose of `BoolConfirmException` is to format a meaningful description that can be read through the `reason` method in the base class.

The next thing we need to do is catch the base class when running the tests and display the confirm reason instead of a message saying that there was an unexpected exception. Modify the `runTests` method in `Test.h` so that it will catch the new exception base class and set the appropriate failed message like this:

```
    try
    {
        test->runEx();
    }
    catch (ConfirmException const & ex)
    {
        test->setFailed(ex.reason());
    }
```

The confirm exception is ready. Building and running shows the following test result:

```
- - - - - - - - - - - - - - -
Test passing grades
Failed
Confirm failed on line 17
    Expected: false
- - - - - - - - - - - - - - -
```

This is a lot better than saying there was an unexpected exception. Now, we understand there was a confirm failure on line 17 and that the test expected the value to be false. Line 17 is for grade 0, which we expected to be a failing grade.

Let's add a macro for the confirm so that we no longer have to provide the line number manually. And the macro can include the backward logic in the `if` condition and the throwing of the proper confirm exception. Here's what the test should look like with the macro. We'll add the macro but only after we write the code that intends to use the macro. Change the passing grades test in `Confirm.cpp` to look like this:

```
TEST("Test passing grades")
{
    bool result = isPassingGrade(0);
```

```
        CONFIRM_FALSE(result);

        result = isPassingGrade(100);
        CONFIRM_TRUE(result);
}
```

Now the test really looks like it's using confirms. Additionally, the macros make it very clear that the first confirm is *expecting* result to be false, while the second confirm is *expecting* result to be true. The value that gets passed to the macro is called the *actual* value. As long as the actual value matches the expected value, then the confirm passes and lets the test continue.

To define these macros, we'll put them at the end of Test.h. Note that each one is almost identical to what the test used to code manually:

```
#define CONFIRM_FALSE( actual ) \
if (actual) \
{ \
    throw MereTDD::BoolConfirmException(false, __LINE__); \
}


#define CONFIRM_TRUE( actual ) \
if (not actual) \
{ \
    throw MereTDD::BoolConfirmException(true, __LINE__); \
}
```

You can see that when confirming a false expected value, the if condition looks for a true actual value. Additionally, when confirming a true expected value, the if condition looks for a false actual value. Both macros throw BoolConfirmException and use __LINE__ to get the line number automatically.

Now, running the tests shows almost the exact same results. The only difference is the line number that the passing grades test fails at. This is because the confirm macros now use a single line each. The results look like this:

```
- - - - - - - - - - - - - - - -
Test passing grades
Failed
Confirm failed on line 15
    Expected: false
- - - - - - - - - - - - - - - -
```

The confirms are much easier to use now, and they make the test clearer to read and understand. Our goal is not to build a school grading application, so we'll be removing the exploratory code. However, before removing it, the next section will use the passing grades test to explain another important aspect of TDD. And that is the question of what to do about error cases.

Should error cases be tested, too?

Is it possible to get to 100% testing code coverage? And what does that mean?

Let me explain by continuing to use the passing grades code we were exploring in the previous section. Here is the test again:

```
TEST("Test passing grades")
{
    bool result = isPassingGrade(0);
    CONFIRM_FALSE(result);

    result = isPassingGrade(100);
    CONFIRM_TRUE(result);
}
```

Right now, this test does cover 100% of the function under test. That means that all of the code inside the `isPassingGrade` function is being run by at least one test. I know, the `isPassingGrade` function is a simple function with a single line of code that always returns true. It looks like this:

```
bool isPassingGrade (int value)
{
    return true;
}
```

With a function this simple, just calling it from within a test will make sure that all of the code is covered or run. As it is, the function doesn't work and needs to be enhanced to pass both confirms. We can enhance it to look like this:

```
bool isPassingGrade (int value)
{
    if (value < 60)
    {
        return false;
    }
```

```
    return true;
}
```

Building and running the project now passes the test. The results of the passing grades test look like this:

```
- - - - - - - - - - - - - - -

Test passing grades
Passed

- - - - - - - - - - - - - -
```

And we still have 100% code coverage for this function because the passing grades test calls the function twice with the values of 0 and 100. The first call causes the `if` condition to be true, which executes the code inside the `if` block. And the second call causes the `return` statement after the `if` block to run. By calling `isPassingGrade` with both the 0 and 100 values, we cause all of the code inside to be run at least once. That is what it means to achieve 100% code coverage.

Both values of 0 and 100 are valid grades, and it makes sense to test with them. We don't need to test what will happen if we call `isPassingGrade` with the values of 1 or 99. That's because they are not interesting.

Edge values are almost always interesting. So, it would make sense to add a couple more calls inside the test for values 59 and 60. While these represent good call values and confirms to add to the test, they won't do anything for the code coverage.

That leads to the first point I want you to understand. Simply achieving 100% code coverage is not enough. You want to ensure that you are testing everything that needs to be tested. Look for edge cases that should be tested even if they don't do anything to improve your code coverage.

And then look for error cases.

Error cases will likely drive your code to add extra checking to make sure the error cases are properly handled. TDD is a great way to drive these conditions. Alternatively, you might decide to change your design as a way to make an error case no longer applicable.

For example, does it make sense to check whether a negative grade is passing? If so, definitely add a test and then add the code to make the test pass. This is something that I would put into a new test. Remember the balance between having a single confirm per test versus allowing multiple confirms?

It makes sense to include all confirms for calling `isPassingGrade` for the values of 0, 59, 60, and 100 in a single test. At least to me.

However, calling `isPassingGrade` with a value of -1 is different enough that it should have its own test. Or maybe thinking of this test is enough to cause you to change the design so that `isPassingGrade` no longer accepts an int parameter, and you decide to use an unsigned int parameter instead. For this particular example, I would probably use an unsigned int. That would mean we no longer need a test for -1 or any negative number grade.

But what about grades above 100? Maybe they should be allowed for extra credit grades. If so, then add a new test for grades above 100 and make sure to confirm they pass. You might find the values of 101, 110, and 1,000,000 to be interesting.

Why the values of 101, 110, and 1,000,000? Well, 101 is an edge value and should be included. The value of 110 seems like a reasonable extra credit value. And the value of 1,000,000 is a good example of a ridiculous value that should be included just to make sure the code doesn't fail with some unexpected exception. You might even consider putting the 1,000,000 value in its own test.

Error cases should be tested. Ideally, you will think of the error cases while writing the tests, and you can write the test first before adding code to handle the error condition. For example, if you decide that any grade over 1,000 should result in an exception being thrown, then write a test that expects the exception and call `isPassingGrade` with the value of 1,000 to make sure that it does throw.

One final thought about testing error cases is this: I've seen a lot of code that was not designed using TDD, and one thing that stands out to me regarding a lot of this code is that error cases are much harder to test. Sometimes, it's no longer feasible to add tests for certain error cases because they are too difficult to isolate and get them to run so that the test can verify how the code responds.

Once you start following TDD, you'll find that you have much better test coverage. That's because you designed tests first, including the tests for error cases. This forces you to make your designs *testable* from the very beginning.

Summary

In this chapter, you learned how to write tests that can detect a failure even before reaching the end of the test. You learned how to use confirms to make sure that the actual values match what you expect them to be. However, this chapter only explained how to check bool values. There are many other types of values you will need to check, such as the following:

- You might have a number such as a count that needs to be confirmed
- You might need to check a string value to make sure it contains the text you expect

The next chapter will add these additional types and explain a common problem when comparing fractional or floating-point numeric values.

5

Adding More Confirm Types

The previous chapter introduced confirmations and showed you how to use them to verify that the bool values within your tests match what you expect them to be. The chapter did this with some exploratory code based on a school grading example. We're going to change the grading example to better fit with a test library and add additional types that you will be able to use in your confirms.

In this chapter, we will cover the following main topics:

- Fixing the bool confirms
- Confirming equality
- Changing the code to fix a problem that line numbers are causing with test failures
- Adding more confirm types
- Confirming string literals
- Confirming floating-point values
- How to write confirms

The additional types add some new twists to confirms that, in this chapter, you'll learn how to work around. By the end of this chapter, you'll be able to write tests that can verify any result you need to be tested.

Technical requirements

All of the code in this chapter uses standard C++ that builds on any modern C++ 17, or later, compiler and standard library. The code is based on and continues from the previous chapters.

You can find all the code for this chapter at the following GitHub repository:

https://github.com/PacktPublishing/Test-Driven-Development-with-CPP

Fixing the bool confirms

The previous chapter explored what it means to confirm a value. However, it left us with some temporary code that we need to fix. Let's start by fixing the code in Confirm.cpp so that it no longer refers to school grades. We want confirms to work with types such as bool. That's why the confirm macros we have now are called CONFIRM_TRUE and CONFIRM_FALSE. The mention of true and false in the macro names are the expected values. Additionally, the macros accept a single parameter, which is the actual value.

Instead of a test about passing grades, let's replace it with a test about bools like this:

```
TEST("Test bool confirms")
{
    bool result = isNegative(0);
    CONFIRM_FALSE(result);

    result = isNegative(-1);
    CONFIRM_TRUE(result);
}
```

The new test is clear about what it tests and needs a new helper function called isNegative instead of the previous function that determined whether a grade was passing or not. I wanted something that is simple and can be used to generate a result with an obvious expected value. The isNegative function replaces the previous isPassingGrade function and looks like this:

```
bool isNegative (int value)
{
    return value < 0;
}
```

This is a simple change that removes the exploratory code based on grades and gives us something that now fits in the test library. Now, in the next section, we can continue with confirms that test for equality.

Confirming equality

In a way, the bool confirms do test for equality. They ensure that the actual bool value is equal to the expected value. This is what the new confirms that are introduced in this chapter will do, too. The only difference is that the CONFIRM_TRUE and CONFIRM_FALSE confirms don't need to accept a parameter for the expected value. Their expected value is implied in their name. We can do this for bool types because there are only two possible values.

However, let's say that we want to verify that an actual int value equals 1. Do we really want a macro that's called CONFIRM_1? We would need billions of macros for each possible 32-bit int and even more for a 64-bit int. And verifying text strings to make sure they match expected values becomes impossible with this approach.

Instead, all we need to do is modify the macros for the other types to accept both an expected value and an actual value. If the two values are not equal to each other, then the macros should result in the test failing with an appropriate message that explains what was expected and what was actually received.

Macros are not designed to resolve different types. They perform simple text replacement only. We'll need real C++ functions to work properly with the different types we'll be checking. Additionally, we might as well change the existing bool macros to call a function instead of defining the code directly inside the macro. Here are the existing bool macros, as we defined them in the previous chapter:

```
#define CONFIRM_FALSE( actual ) \
if (actual) \
{ \
    throw MereTDD::BoolConfirmException(false, __LINE__); \
}


#define CONFIRM_TRUE( actual ) \
if (not actual) \
{ \
    throw MereTDD::BoolConfirmException(true, __LINE__); \
}
```

What we need to do is move the if and throw statements into a function. We only need one function for both true and false, and it will look like this:

```
inline void confirm (
    bool expected,
    bool actual,
    int line)
{
    if (actual != expected)
    {
        throw BoolConfirmException(expected, line);
    }
}
```

This function can be placed in Test.h inside the MereTDD namespace right before TestBase is defined. The function needs to be inline and no longer needs to qualify the exception with the namespace since it's now inside the same namespace.

Also, you can see better that this is an equality comparison even for bool values. The function checks to make sure that the actual value is equal to the expected value, and if not, then it throws an exception. The macros can be simplified to call the new function like this:

```
#define CONFIRM_FALSE( actual ) \
    MereTDD::confirm(false, actual, __LINE__)
#define CONFIRM_TRUE( actual ) \
    MereTDD:: confirm(true, actual, __LINE__)
```

Building and running show that all of the tests pass, and we are ready to add additional types to confirm. Let's start with a new test in Confirm.cpp for int types like this:

```
TEST("Test int confirms")
{
    int result = multiplyBy2(0);
    CONFIRM(0, result);

    result = multiplyBy2(1);
    CONFIRM(2, result);

    result = multiplyBy2(-1);
    CONFIRM(-2, result);
}
```

Instead of a bools, this code tests int values. It uses a new helper function that should be simple to understand, which just multiplies a value by 2. We need the new helper function to be declared at the top of the same file like this:

```
int multiplyBy2 (int value)
{
    return value * 2;
}
```

The test won't build yet. That's okay because, when using a TDD approach, we want to focus on the usage first. This usage seems good. It will let us confirm that any int value is equal to whatever we expect it to be. Let's create the CONFIRM macro and place it right after the two existing macros that confirm true and false like this:

```
#define CONFIRM_FALSE( actual ) \
    MereTDD::confirm(false, actual, __LINE__)
#define CONFIRM_TRUE( actual ) \
    MereTDD:: confirm(true, actual, __LINE__)
#define CONFIRM( expected, actual ) \
    MereTDD::confirm(expected, actual, __LINE__)
```

Changing the macros to call a function is really paying off now. The CONFIRM macro needs an extra parameter for the expected value, and it can call the same function name. How can it call the same function, though? Well, that's because we're going to overload the function. What we have now only works for bool values. This is why we switched to a design that can make use of data types. All we need to do is provide another implementation of confirm that is overloaded to work with ints like this:

```
inline void confirm (
    int expected,
    int actual,
    int line)
{
    if (actual != expected)
    {
        throw ActualConfirmException(expected, actual, line);
    }
}
```

This is almost identical to the existing confirm function. It takes ints for the expected and actual parameters instead of bools and will throw a new exception type. The reason for the new exception type is so that we can format a failure message that will display both the expected and actual values. The BoolConfirmException type will only be used for bools and will format a message that only mentions what was expected. Additionally, the new ActualConfirmException type will format a message that mentions both the expected and actual values.

The new exception type looks like this:

```
class ActualConfirmException : public ConfirmException
{
public:
    ActualConfirmException (int expected, int actual, int line)
    : mExpected(std::to_string(expected)),
```

```
        mActual(std::to_string(actual)),
        mLine(line)
    {
        formatReason();
    }

private:
    void formatReason ()
    {
        mReason =  "Confirm failed on line ";
        mReason += std::to_string(mLine) + "\n";
        mReason += "    Expected: " + mExpected + "\n";
        mReason += "    Actual  : " + mActual;
    }

    std::string mExpected;
    std::string mActual;
    int mLine;
};
```

You might be wondering why the new exception type stores the expected and actual values as strings. The constructor accepts ints and then converts the ints into strings before formatting the reason. This is because we'll be adding multiple data types, and we don't really need to do anything different. Each type just needs to display a descriptive message based on strings when a test fails.

We don't need to use the expected or actual values for any calculations. They just need to be formatted into a readable message. Additionally, this design will let us use a single exception for all the data types other than bool. We could use this new exception for bools too, but the message doesn't need to mention the actual value for bools. So, we'll keep the existing exception for bools and use this new exception type for everything else.

By storing the expected and actual values as strings, all we need is an overloaded constructor for each new data type we want to support. Each constructor can convert the expected and actual values into strings that can then be formatted into a readable message. This is better than having an IntActualConfirmException class, a StringActualConfirmException class, and more.

We can build and run the tests again. The results for both the bool and int tests look like this:

```
- - - - - - - - - - - - - - -
Test bool confirms
```

```
Passed
----------------
Test int confirms
Passed
----------------
```

So, what happens if the confirms fail? Well, we've already seen in the previous chapter what a failed bool confirm looks like. But we don't yet have any tests for failure cases. We should add them and make them expected failures so that the behavior can be captured. Even a failure should be tested to make sure it remains a failure. It would be bad if, in the future, we made some changes to the code that turned a failure into a success. That would be a breaking change because a failure should be expected. Let's add a couple of new tests to Confirm.cpp like this:

```
TEST("Test bool confirm failure")
{
    bool result = isNegative(0);
    CONFIRM_TRUE(result);
}

TEST("Test int confirm failure")
{
    int result = multiplyBy2(1);
    CONFIRM(0, result);
}
```

We get the expected failures, and they look like this:

```
----------------
Test bool confirm failure
Failed
Confirm failed on line 41
    Expected: true
----------------
Test int confirm failure
Failed
Confirm failed on line 47
    Expected: 0
    Actual  : 2
----------------
```

The next step is to set the expected failure messages so that these tests pass instead of fail. However, there's a problem. The line number is part of the error message. We want the line number to be displayed in the test results. But that means we also have to include the line number in the expected failure message in order to treat the failures as passing. Why is this a problem? Well, that's because every time a test is moved or even when other tests are added or removed, the line numbers will change. We don't want to have to change the expected error messages for something that is not really part of the error. The line number tells us where the error happened and should not be part of the reason for why it happened.

In the next section, we'll fix the line numbers with some refactoring.

Decoupling test failures from line numbers

We need to remove the line number from the confirm failure reason so that tests can be given an expected failure reason that won't change as the test is moved or shifted to different locations in the source code file.

This type of change is called *refactoring*. We're not going to make any changes that cause different or new behaviors to appear in the code. At least, that's the goal. Using TDD will help you refactor your code because you should already have tests in place for all of the important aspects.

Refactoring with proper tests lets you verify that nothing has changed. Many times, refactoring without TDD is avoided because the risk of introducing new bugs is too great. This tends to make problems bigger, as the refactoring is delayed or avoided entirely.

We have a problem with the line numbers. We could ignore the problem and just update the tests with new line numbers in the expected failure messages anytime a change is made. But that is not right and will only lead to more work and brittle tests. As more tests are added, the problem will only get worse. We really should fix the problem now. Because we're following TDD, we can feel confident that the changes we are about to make will not break anything that has already been tested. Or, at least, if it does break, we'll know about it and can fix any breaks right away.

The first step is to add line number information to the `ConfirmException` base class in `Test.cpp`:

```
class ConfirmException
{
public:
    ConfirmException (int line)
    : mLine(line)
    { }
```

```
        virtual ~ConfirmException () = default;

        std::string_view reason () const
        {
            return mReason;
        }

        int line () const
        {
            return mLine;
        }

    protected:
        std::string mReason;
        int mLine;
    };
```

Then, in the runTests function, we can get the line from the confirm exception and use it to set the failure location in the test like this:

```
        try
        {
            test->runEx ();
        }
        catch (ConfirmException const & ex)
        {
            test->setFailed (ex.reason (), ex.line ());
        }
```

Even though we are not starting with a test, notice how I'm still following a TDD approach to writing the code, as I'd like it to be used before implementing it fully. This is a great example because I originally thought about adding a new method to the test class. It was called setFailedLocation. But that made the existing setFailed method seem strange. I almost renamed setFailed to setFailedReason, which would have meant that it would need to be changed in the other places it's called. Instead, I decided to add an extra parameter for the line number to the existing setFailed method. I also decided to give the parameter a default value so that the other code would not need to be changed. This makes sense and lets the caller set the failed reason by itself or with a line number if known.

We need to add a line number data member to the `TestBase` class. The line number will only be known for confirms, so it will be called `mConfirmLocation` like this:

```cpp
    std::string mName;
    bool mPassed;
    std::string mReason;
    std::string mExpectedReason;
    int mConfirmLocation;
};
```

The new data member needs to be initialized in the `TestBase` constructor. We'll use the value of -1 to mean that the line number location is not applicable:

```cpp
    TestBase (std::string_view name)
    : mName(name), mPassed(true), mConfirmLocation(-1)
    { }
```

We need to add the line number parameter to the `setFailed` method like this:

```cpp
    void setFailed (std::string_view reason,
        int confirmLocation = -1)
    {
        mPassed = false;
        mReason = reason;
        mConfirmLocation = confirmLocation;
    }
```

Additionally, we need to add a new getter method for the confirm location like this:

```cpp
    int confirmLocation () const
    {
        return mConfirmLocation;
    }
```

This will let the `runTests` function set the line number when it catches a confirm exception, and the test will be able to remember the line number. At the end of `runTests`, where the failure message is sent to the output, we need to test `confirmLocation` and change the output if we have a line number or not, as follows:

```cpp
        else
        {
```

```
            ++numFailed;
            if (test->confirmLocation() != -1)
            {
                output << "Failed confirm on line "
                    << test->confirmLocation() << "\n";
            }
            else
            {
                output << "Failed\n";
            }
            output << test->reason()
                << std::endl;
        }
```

This will also fix a minor problem with confirms. Previously, the test results printed a line that said the test failed and then another line that said a confirm failed. The new code will only display either a generic failed message or a confirm failed message with a line number.

We're not done yet. We need to change both derived exception class constructors to initialize the base class line number and to stop including the line number as part of the reason. The constructor for `BoolConfirmException` looks like this:

```
    BoolConfirmException (bool expected, int line)
    : ConfirmException(line)
    {
        mReason += "    Expected: ";
        mReason += expected ? "true" : "false";
    }
```

Additionally, the `ActualConfirmException` class needs to be changed throughout. The constructor needs to initialize the base class with the line number, the formatting needs to change, and the line number data member can be removed since it's now in the base class. The class looks like this:

```
  class ActualConfirmException : public ConfirmException
  {
  public:
      ActualConfirmException (int expected, int actual, int line)
      : ConfirmException(line),
        mExpected(std::to_string(expected)),
```

```
        mActual(std::to_string(actual))
    {
        formatReason();
    }

private:
    void formatReason ()
    {
        mReason += "    Expected: " + mExpected + "\n";
        mReason += "    Actual   : " + mActual;
    }

    std::string mExpected;
    std::string mActual;
};
```

We can build again and running still shows the expected failures. The failure reasons are formatted slightly differently than before and look like this:

```
--------------
Test bool confirm failure
Failed confirm on line 41
    Expected: true
--------------
Test int confirm failure
Failed confirm on line 47
    Expected: 0
    Actual   : 2
--------------
```

It looks almost the same, which is good. Now we can set the expected failure messages without needing to worry about the line numbers like this:

```
TEST("Test bool confirm failure")
{
    std::string reason = "    Expected: true";
    setExpectedFailureReason(reason);
```

```
    bool result = isNegative(0);
    CONFIRM_TRUE(result);
}


TEST("Test int confirm failure")
{
    std::string reason = "    Expected: 0\n";
    reason += "    Actual  : 2";
    setExpectedFailureReason(reason);

    int result = multiplyBy2(1);
    CONFIRM(0, result);
}
```

Notice that the expected failure reason needs to be formatted to exactly match what the test displays when it fails. This includes spaces used to indent and new lines. Once the expected failure reasons are set, all of the tests pass again like this:

```
- - - - - - - - - - - - - -
Test bool confirm failure
Expected failure
    Expected: true
- - - - - - - - - - - - - -
Test int confirm failure
Expected failure
    Expected: 0
    Actual  : 2
- - - - - - - - - - - - - -
```

Both tests are expected failures and are treated as passing. Now we can continue adding more confirm types.

Adding more confirm types

Currently, we can confirm bool and int values inside the tests. We need more than this, so what should we add next? Let's add support for the long type. It's similar to an int and, on many platforms, will effectively be the same. Even though it may or may not use the same number of bits as an int, to the

C++ compiler, it is a different type. We can begin by adding a basic test to `Confirm.cpp` that tests the long type like this:

```
TEST("Test long comfirms")
{
    long result = multiplyBy2(0L);
    CONFIRM(0L, result);

    result = multiplyBy2(1L);
    CONFIRM(2L, result);

    result = multiplyBy2(-1L);
    CONFIRM(-2L, result);
}
```

The test calls the same `multiplyBy2` helper function, which performs extra conversions because it's not working with longs throughout. We start with long literal values by adding the L suffix. These get converted into ints in order to be passed to `multiplyBy2`. The return value is also an int, which gets converted into a long in order to be assigned to `result`. Let's prevent all of this extra conversion by creating an overloaded version of `multiplyBy2` that accepts a long type and returns a long type:

```
long multiplyBy2 (long value)
{
    return value * 2L;
}
```

If we try to build right now, there will be an error because the compiler doesn't know which overload of `confirm` to call. The only available choices are to either convert the long expected and actual values into ints or bools. Neither choice is a match, and the compiler sees the call as ambiguous. Remember that the CONFIRM macro gets transformed into a call to the overloaded `confirm` function.

We can fix this by adding a new overloaded version of `confirm` that uses long parameters. However, a better solution is to change the existing version of `confirm` that uses int parameters into a template like this:

```
template <typename T>
void confirm (
    T const & expected,
    T const, & actual,
    int line)
```

```
{
    if (actual != expected)
    {
        throw ActualConfirmException(
            std::to_string(expected),
            std::to_string(actual),
            line);
    }
}
```

We still have the version of confirm that uses a bool parameter. The template will match both int and long types. Additionally, the template will match types that we don't yet have tests for. The new templated confirm method also does the conversion into std::string when creating the exception to be thrown. In *Chapter 12, Creating Better Test Confirmations*, you'll see that there is a problem with how we convert the expected and actual values into strings. Or, at least, there is a better way. What we have does work but only for numeric types that can be passed to std::to_string.

Let's update the ActualConfirmException constructor to use strings that we will now be calling std::to_string from within the confirm function. The constructor looks like this:

```
ActualConfirmException (
    std::string_view expected,
    std::string_view actual,
    int line)
: ConfirmException(line),
  mExpected(expected),
  mActual(actual)
{
    formatReason();
}
```

Everything builds, and all the tests pass again. We can add a new test in Confirm.cpp for a long failure like this:

```
TEST("Test long confirm failure")
{
    std::string reason = "    Expected: 0\n";
    reason += "    Actual   : 2";
    setExpectedFailureReason(reason);
```

```
    long result = multiplyBy2(1L);
    CONFIRM(0L, result);
}
```

The failure reason string is the same as for an int even though we are testing a long type. The test result for the new test looks like this:

```
- - - - - - - - - - - - - - -
Test long confirm failure
Expected failure
    Expected: 0
    Actual  : 2
- - - - - - - - - - - - - - -
```

Let's try a type that will show something different. A long long type can definitely hold numeric values that are bigger than an int. Here is a new test in Confirm.cpp that tests long long values:

```
TEST("Test long long confirms")
{
    long long result = multiplyBy2(0LL);
    CONFIRM(0LL, result);

    result = multiplyBy2(10'000'000'000LL);
    CONFIRM(20'000'000'000LL, result);

    result = multiplyBy2(-10'000'000'000LL);
    CONFIRM(-20'000'000'000LL, result);
}
```

With a long long type, we can have values greater than a maximum 32-bit signed value. The code uses single quote marks to make the larger numbers easier to read. The compiler ignores the single quote marks, but they help us to visually separate every group of thousands. Also, the suffix, LL, tells the compiler to treat the literal value as a long long type.

The result for this passing test looks like the others:

```
- - - - - - - - - - - - - - -
Test long long confirms
Passed
- - - - - - - - - - - - - - -
```

We need to look at a `long long` failure test result to see the larger numbers. Here is a failure test:

```
TEST("Test long long confirm failure")
{
    std::string reason = "    Expected: 10000000000\n";
    reason += "    Actual   : 20000000000";
    setExpectedFailureReason(reason);

    long long result = multiplyBy2(10'000'000'000LL);
    CONFIRM(10'000'000'000LL, result);
}
```

Because we're not formatting the output with separators, we need to use the unadorned numbers in text format without any commas. This is probably for the best anyway because some locales use commas and some use dots. Note that we don't try to do any formatting, so the expected failure message also uses no formatting.

Now we can see that the failure description does indeed match the larger numbers and looks like this:

```
- - - - - - - - - - - - - - -
Test long long confirm failure
Expected failure
    Expected: 10000000000
    Actual   : 20000000000
- - - - - - - - - - - - - - -
```

I want to highlight one important point about failure tests. They are purposefully using incorrect expected values to force a failure. You will not do this in your tests. But then you also will not need to write tests that you *want* to fail. We want these tests to fail so that we can verify that the test library is able to properly detect and handle any failures. Because of this, we treat the failures as passes.

We could keep going and add tests for shorts, chars, and all of the unsigned versions. However, this is becoming uninteresting at this point because all we are doing is testing that the template function works properly. Instead, let's focus on types that use non-template code that has been written to work properly.

Here is a simple test for the string type:

```
TEST("Test string confirms")
{
    std::string result = "abc";
    std::string expected = "abc";
```

```
        CONFIRM(expected, result);
}
```

Instead of writing a fake helper method that returns a string, this test simply declares two strings and will use one as the actual value and the other as the expected value. By initializing both strings with the same text, we expect them to be equal, so we call CONFIRM to make sure they are equal.

When you are writing a test, you will want to assign result a value that you get from the function or method that you are testing. Our goal here is to test that the CONFIRM macro and the underlying test library code work properly. So, we can skip the function being tested and go straight to the macro with two string values where we know what to expect.

This seems like a reasonable test. And it is. But it doesn't compile. The problem is that the confirm template function tries to call std::to_string on the values provided. This doesn't make sense when the values are already strings.

What we need is a new overload of confirm that uses strings. We'll actually create two overloads, one for string views and one for strings. The first overload function looks like this:

```
inline void confirm (
    std::string_view expected,
    std::string_view actual,
    int line)
{
    if (actual != expected)
    {
        throw ActualConfirmException(
            expected,
            actual,
            line);
    }
}
```

This first function takes string views, which will be a better match than the template method when working with string views. Then, it passes the strings given to the ActualConfirmException constructor without trying to call std::to_string because they are already strings.

The second overloaded function looks like this:

```
inline void confirm (
    std::string const & expected,
    std::string const & actual,
```

```
        int line)
{
    confirm(
        std::string_view(expected),
        std::string_view(actual),
        line);
}
```

This second function takes constant string references, which will also be a better match than the template method when working with strings. Then, it converts the strings into string views and calls the first function.

Now we can add a string failure test like this:

```
TEST("Test string confirm failure")
{
    std::string reason = "    Expected: def\n";
    reason += "    Actual   : abc";
    setExpectedFailureReason(reason);

    std::string result = "abc";
    std::string expected = "def";
    CONFIRM(expected, result);
}
```

The test result after building and running the tests looks like this:

```
---------------
Test string confirm failure
Expected failure
    Expected: def
    Actual   : abc
---------------
```

There's one more important aspect to consider about strings. We need to consider string literals that are really constant char pointers. We'll explore pointers followed by string literals in the next section.

Confirming string literals

A string literal might look like a string, but the C++ compiler treats a string literal as a pointer to the first of a set of constant chars. The set of constant chars is terminated with a null character value, which is the numeric value of zero. That's how the compiler knows how long the string is. It just keeps going until it finds the null. The reason that the chars are constant is that the data is normally stored in memory that is write protected, so it cannot be modified.

When we try to confirm a string literal, the compiler sees a pointer and has to decide which overloaded confirm function to call. Before we get too far with our exploration of string literals, what other problems can we get into with pointers?

Let's start with the simple bool type and see what kinds of problems we run into if we try to confirm bool pointers. This will help you to understand string literal pointers by, first, understanding a simpler example test for bool pointers. You don't need to add this test to the project. It is included here just to explain what happens when we try to confirm a pointer. The test looks like this:

```
TEST("Test bool pointer confirms")
{
    bool result1 = true;
    bool result2 = false;
    bool * pResult1 = &result1;
    bool * pResult2 = &result2;
    CONFIRM_TRUE(pResult1);
    CONFIRM_FALSE(pResult2);
}
```

The preceding test actually compiles and runs. But it fails with the following result:

```
---------------
Test bool pointer confirms
Failed confirm on line 86
    Expected: false
---------------
```

Line 86 is the second confirm in the test. So, what is going on? Why does the confirm think that pResult2 points to a true value?

Well, remember that the confirm macro just gets replaced with a call to one of the confirm methods. The second confirm deals with the following macro:

```
#define CONFIRM_FALSE( actual ) \
    confirm(false, actual, __LINE__)
```

And it tries to call `confirm` with a hardcoded false bool value, the bool pointer that was passed to the macro, and the int line number. There is no exact match for a bool, a bool pointer, or an int for any version of `confirm`, so something either has to be converted or the compiler will generate an error. We know there was no error because the code compiled and ran. So, what got converted?

This is a great example of the TDD process, as explained in *Chapter 3, The TDD Process*, to write the code first as you want it to be used and compile it even if you expect the build to fail. In this case, the build did not fail, and that gives us insight that we might have otherwise missed.

The compiler was able to convert the pointer value into a bool and that was seen as the best choice available. In fact, I didn't even get a warning about the conversion. The compiler silently made the decision to convert the pointer to a bool into a bool value. This is almost never what you want to happen.

So, what does it even mean to convert a pointer into a bool? Any pointer with a valid nonzero address will get converted into true. Additionally, any null pointer with a zero address will get converted into false. Because we have the real address of `result2` stored in the `pResult2` pointer, the conversion was made to a true bool value.

You might be wondering what happened to the first confirm and why it did not fail. Why did the test proceed to the second confirm before it failed? Well, the first confirm went through the same conversion for a bool, bool pointer, and int. Both conversions resulted in a true bool value because both pointers held valid addresses.

The first confirm called `confirm` with true, true, and the line number, which passed. But the second confirm called `confirm` with false, true, and the line number, which failed.

To solve this, we either need to add support for pointers of all types or remember to dereference the pointers before confirming them. Adding support for pointers might seem like a simple solution until we get to string literals, which are also pointers. It's not as simple as it seems and is not something we need to do now. Let's keep the test library as simple as possible. Here is how you can fix the bool confirm test shown earlier:

```
TEST("Test bool pointer dereference confirms")
{
    bool result1 = true;
    bool result2 = false;
    bool * pResult1 = &result1;
    bool * pResult2 = &result2;
    CONFIRM_TRUE(*pResult1);
    CONFIRM_FALSE(*pResult2);
}
```

Notice that the tests dereference the pointers instead of passing the pointers directly to the macros. This means that the test is really just testing bool values, and that's why I said that you really don't need to add the test.

String literals are frequently found in the source code. They are an easy way to represent an expected string value. The problem with string literals is they are not strings. They are a pointer to a constant char. And we can't just dereference a string literal pointer as we did for a bool pointer. That would result in a single char. We want to confirm the whole string.

Here is a test that shows what will likely be the major usage of string literals. The most common usage will be comparing a string literal with a string. The test looks like this:

```
TEST("Test string and string literal confirms")
{
    std::string result = "abc";
    CONFIRM("abc", result);
}
```

This works because one of the argument types that ends up getting passed to the confirm function is std::string. The compiler doesn't find an exact match for both arguments; however, because one is a string, it decides to convert the string literal into a string, too.

Where we run into problems is when we try to confirm two string literals for both the expected and actual values. The compiler sees two pointers and has no clue that they should both be converted into strings. This is not a normal situation that you will need to verify in a test. Additionally, if you ever do need to compare two string literals, it's easy to wrap one of them into a std::string argument type before confirming.

Also, in *Chapter 12*, *Creating Better Test Confirmations*, you'll see how you can get around the problem of confirming two string literals. We'll be improving the whole design used to confirm the test results. The design we have now is often called the classic way to confirm values. *Chapter 12* will introduce a new way that is more extensible, easier to read, and more flexible.

We've come a long way in terms of adding support for different types, and you also understand how to work with string literals. However, I've stayed away from the two floating-point types, float and double, because they need some special consideration. They will be explained next.

Confirming floating point values

At the most basic level, confirms work by comparing an expected value with an actual value and throwing an exception if they are different. This works for all the integral types such as int and long, bool types, and even strings. The values either match or don't match.

This is where things get difficult for the float and double floating point types – because it's not always possible to accurately compare two floating-point values.

Even in the decimal system that we are used to from grade school, we understand there are some fractional values that can't be accurately represented. A value such as 1/3 is easy to represent as a fraction. But writing it in a floating-point decimal format looks like 0.33333 with the digit 3 continuing forever. We can get close to the true value of 1/3, but at some point, we have to stop when writing 0.333333333... And no matter how many 3s we include, there are always more.

In C++, floating-point values use a binary number system that has similar accuracy issues. But the accuracy issues in binary are even more common than in decimal.

I won't go into all the details because they are not important. However, the main cause of the extra issues in binary is caused by there being fewer factors of 2 than there are of 10. With the base 10 decimal system, the factors are 1, 2, 5, and 10. While in binary, the factors of 2 are only 1 and 2.

So, why are the factors important? Well, it's because they determine which fractions can be accurately described and which cannot. A fraction such as 1/3 causes trouble for both systems because 3 is not a factor in either. Another example is 1/7. These fractions are not very common, though. The fraction of 1/10 is very common in decimal. Because 10 is a factor, this means that values such as 0.1, 0.2, 0.3, and more can all be accurately represented in decimal.

Additionally, because 10 is not a factor in binary base 2, these same values that are widely used have no representation with a fixed number of digits as they do in decimal.

So, what all of this means is that if you have a binary floating-point value that looks like 0.1, it is close to the actual value but can't quite be exact. It might be displayed as 0.1 when converted into a string but that also involves a little bit of rounding.

Normally, we don't worry about the computer's inability to accurately represent values that we are used to being exact from grade school – that is, until we need to test one floating-point value to see if it equals another.

Even something as simple as 0.1 + 0.2 that looks like 0.3 will probably not equal 0.3.

When comparing computer floating-point values, we always have to allow for a certain amount of error. As long as the values are close, we can assume they are equal.

However, the ultimate problem is that there is no good single solution that can determine whether two values are close. The amount of error we can represent changes depending on how big or how small the values are. Floating-point values change drastically when they get really close to 0. And they lose the ability to represent small values as they get larger. Because floating-point values can get really large, the amount of accuracy that is lost with large values can also be large.

Let's imagine if a bank used floating-point values to keep track of your money. Would you be happy if your bank could no longer track anything less than a thousand dollars just because you have billions? We're no longer talking about losing a few cents. Or maybe you only have 30 cents in your account and you want to withdraw all 30 cents. Would you expect the bank to deny your withdrawal because it thinks 30 cents is more than the 30 cents you have? These are the types of problems that floating-point values can lead to.

Because we're following a TDD process, we're going to start out simple with floating point values and include a small margin of error when comparing either float, double, or long double values to see whether they are equal. We're not going to get fancy and try to adjust the margin depending on how big or small the values are.

Here is a test that we will use for the float values:

```
TEST("Test float confirms")
{
    float f1 = 0.1f;
    float f2 = 0.2f;
    float sum = f1 + f2;
    float expected = 0.3f;
    CONFIRM(expected, sum);
}
```

The test for float types actually passes on my computer.

So, what happens if we create another test for double types? The new double test looks like this:

```
TEST("Test double confirms")
{
    double d1 = 0.1;
    double d2 = 0.2;
    double sum = d1 + d2;
    double expected = 0.3;
    CONFIRM(expected, sum);
}
```

This test is almost identical, yet it fails on my computer. And the crazy part is that the failure description makes no sense unless you understand that values can be printed as text, which has been adjusted to appear like a nice round number when it really is not. Here is what the failure message shows on my computer:

```
---------------
Test double confirms
Failed confirm on line 122
    Expected: 0.300000
    Actual  : 0.300000
---------------
```

Looking at the message, you might ask how it is possible that 0.300000 does not equal 0.300000. The reason is that neither the expected nor the actual values are exactly 0.300000. They have both been adjusted slightly so that they will display these round-looking values.

The test for long doubles is almost the same as for doubles. Only the types have been changed, as follows:

```
TEST("Test long double confirms")
{
    long double ld1 = 0.1;
    long double ld2 = 0.2;
    long double sum = ld1 + ld2;
    long double expected = 0.3;
    CONFIRM(expected, sum);
}
```

The long double test also fails on my machine for the same reason as the test with doubles. We can fix all of the floating-point confirms by adding special overloads for all three of these types.

Here is an overloaded confirm function that uses a small margin of error when comparing float values:

```
inline void confirm (
    float expected,
    float actual,
    int line)
{
    if (actual < (expected - 0.0001f) ||
        actual > (expected + 0.0001f))
    {
        throw ActualConfirmException(
            std::to_string(expected),
            std::to_string(actual),
            line);
    }
}
```

We need almost the same overload for doubles as for floats. Here is the double overload that does the comparison with a margin of error that is plus or minus the expected value:

```
inline void confirm (
    double expected,
```

```
        double actual,
        int line)
    {
        if (actual < (expected - 0.000001) ||
            actual > (expected + 0.000001))
        {
            throw ActualConfirmException(
                std::to_string(expected),
                std::to_string(actual),
                line);
        }
    }
```

Other than the type changes from float to double, this method uses a smaller margin of error and leaves off the f suffix from the literal values.

The overload function for long doubles is similar to the one for doubles, as follows:

```
inline void confirm (
    long double expected,
    long double actual,
    int line)
{
    if (actual < (expected - 0.000001) ||
        actual > (expected + 0.000001))
    {
        throw ActualConfirmException(
            std::to_string(expected),
            std::to_string(actual),
            line);
    }
}
```

After adding these overloads for floats, doubles, and long doubles, all the tests pass again. We'll be revisiting the problem of comparing floating-point values again in *Chapter 13, How to Test Floating-Point and Custom Values*. The comparison solution we have is simple and will work for now.

We have also covered all of the confirm types we'll be supporting at this time. Remember the TDD rule to do only what is necessary. We can always enhance the design of confirmations later, and that's exactly what we'll be doing in *Chapter 12, Creating Better Test Confirmations*.

Before ending this chapter, I have some advice on writing confirmations. It's nothing that we haven't already been doing, but it does deserve mentioning so that you are aware of the pattern.

How to write confirms

Usually, there are many different ways you can write your code and your tests. What I'll share here is based on years of experience, and while it's not the only way to write tests, I hope you learn from it and follow a similar style. Specifically, I want to share guidance on how to write confirms.

The most important thing to remember is to keep your confirms outside of the normal flow of your tests but still close to where they are needed. When a test runs, it performs various activities that you want to ensure work as expected. You can add confirms along the way to make sure the test is making progress as expected. Or maybe you have a simple test that does one thing and needs one or more confirms at the end to make sure everything worked. All of this is good.

Consider the following three examples of test cases. They each do the same thing, but I want you to focus on how they are written. Here is the first example:

```
TEST("Test int confirms")
{
    int result = multiplyBy2(0);
    CONFIRM(0, result);

    result = multiplyBy2(1);
    CONFIRM(2, result);

    result = multiplyBy2(-1);
    CONFIRM(-2, result);
}
```

This test is one that was used earlier to make sure that we can confirm int values. Notice how it performs an action and assigns the result to a local variable. Then, that variable is checked to make sure its value matches what is expected. If so, the test proceeds to perform another action and assign the result to the local variable. This pattern continues, and if all the confirms match the expected values, the test passes.

Here is the same test written in a different form:

```
TEST("Test int confirms")
{
    CONFIRM(0, multiplyBy2(0));
```

```
        CONFIRM(2, multiplyBy2(1));


        CONFIRM(-2, multiplyBy2(-1));
    }
```

This time, there is no local variable to store the result of each action. Some people would consider this an improvement. It *is* shorter. But I feel this hides what is being tested. I find it better to think about confirms as something that can be removed from a test without changing what a test does. Of course, if you do remove a confirm, then the test might miss a problem that the confirm would have caught. I'm talking about mentally ignoring confirms to get a feel for what a test does, and then thinking about what makes sense to verify along the way. Those verification points become the confirms.

Here is another example:

```
    TEST("Test int confirms")
    {
        int result1 = multiplyBy2(0);

        int result2 = multiplyBy2(1);

        int result3 = multiplyBy2(-1);

        CONFIRM(0, result1);
        CONFIRM(2, result2);
        CONFIRM(-2, result3);
    }
```

This example avoids putting the test steps inside the confirms. However, I feel that it goes too far to separate the test steps from the confirms. There's nothing wrong with sprinkling confirms into your test steps. Doing so lets you catch problems right away. This example puts all the confirms at the end, which means that it also has to wait until the end to catch any problems.

And then there's the problem of the multiple result variables needed so that each can be checked later. This code looks too forced to me – like a programmer who took the long way to reach a goal when there was a simple path available instead.

The first example shows the style of tests written so far in this book, and now you can see why they have been written in this manner. They use confirms where needed and as close to the point of verification as possible. And they avoid placing actual test steps inside the confirms.

Summary

This chapter took us past the simple ability to confirm true and false values. You can now verify anything you need to make sure it matches what you expect.

We simplified the confirm macros by putting the code into overloaded functions with a templated version to handle other types. You saw how to confirm simple data types and work with pointers by dereferencing them first.

The code needed to be refactored, and you saw how TDD helps when you need to make design changes to your code. I could have written the code in this book to make it seem like the code was written perfectly from the very beginning. But that wouldn't help you because nobody writes perfect code from the beginning. As our understanding grows, we sometimes need to change the code. And TDD gives you the confidence to make those changes as soon as they become known instead of waiting – because problems that you delay have a tendency to get bigger instead of going away.

And you should be gaining an understanding of how to write your tests and the best way to incorporate confirms into your tests.

Up until now, we've been working with the C++ features and capabilities found in C++ 17. There is an important new feature found in C++ 20 that will help us get line numbers from the compiler. The next chapter will add this C++ 20 feature and explore some alternate designs. Even if we stay with the same overall design we have now, the next chapter will help you to understand how other testing libraries might do things differently.

6

Explore Improvements Early

We've come a long way with the testing library and have been using TDD the entire time to get us here. Sometimes, it's important to explore new ideas before a project gets too far. After creating anything, we'll have insights that we didn't have at the beginning. And after working with a design for a while, we'll develop a feel for what we like and what we might want to change. I encourage you to take this time to reflect on a design before proceeding.

We have something that is working and a bit of experience using it, so is there anything that we can improve?

This approach is like a higher-level process of TDD, as explained in *Chapter 3*, *The TDD Process*. First, we work out how we'd like to use something, then get it built, then do the minimal amount of work to get it working and the tests passing, and then enhance the design. We've got many things working now, but we haven't gone so far yet where it would be too hard to change. We're going to look at ways that the overall design could be enhanced.

At this point, it's also a good idea to look around at other similar solutions and compare them. Get ideas. And try some new things to see whether they might be better. I've done this and would like to explore two topics in this chapter:

- Can we use a new feature of C++ 20 to get line numbers instead of using __LINE__?
- What would the tests look like if we used lambdas?

By the end of this chapter, you'll understand the importance of and the process involved in exploring improvements early on in the design of your projects. Even if you don't always decide to accept new ideas and make changes, your project will be better because you have taken the time to consider alternatives.

Technical requirements

The code in this chapter uses standard C++, and we will try out a feature introduced in C++ 20. The code is based on and continues from the previous chapters.

You can find all of the code for this chapter at the following GitHub repository:

`https://github.com/PacktPublishing/Test-Driven-Development-with-CPP`

Getting line numbers without macros

C++ 20 includes a new class that will help us get line numbers. In fact, it has a lot more information than just the line number. It includes the name of the file, the function name, and even the column number. However, we only need the line number. Note that at the time of writing this book, the implementation of this new class for my compiler has a bug. The end result is that I have had to put the code back to the way it was before the changes described in this section.

The new class is called `source_location`, and once it finally works correctly, we can change all of the existing `confirm` functions so that they accept `std::source_location` instead of the int for the line number. One example of an existing `confirm` function looks like this:

```
inline void confirm (
    bool expected,
    bool actual,
    int line)
{
    if (actual != expected)
    {
        throw BoolConfirmException(expected, line);
    }
}
```

We can eventually update the confirm functions to use `std::source_location` by changing all of the `confirm` functions, including the template override, to be similar to the following:

```
inline void confirm (
    bool expected,
    bool actual,
    const std::source_location location =
        std::source_location::current())
{
    if (actual != expected)
    {
        throw BoolConfirmException(expected, location.line());
    }
}
```

We're not going to be making these changes right now because of the bug. The code does work as long as there is only a single source file in the project that tries to use `source_location`.

The moment more than one source file tries to use source_location, there is a linker warning and the line method returns bad data. The bug should eventually get fixed, and I'm leaving this section in the book because it is a better approach. Depending on what compiler you are using, you might be able to start using source_location now.

Not only does the last parameter type and name change, but the usage needs to change when the line number is passed to the exception when it's thrown. Notice how the new parameter includes a default value that gets set to the current location. The default parameter value means we no longer need to pass anything for the line number. The new location will get a default value that includes the current line number.

We need to include the header file for source_location at the top of Test.h, as follows:

```
#include <ostream>
#include <source_location>
#include <string_view>
#include <vector>
```

The macros that call confirm need to be updated to no longer worry about the line number:

```
#define CONFIRM_FALSE( actual ) \
    MereTDD::confirm(false, actual)
#define CONFIRM_TRUE( actual ) \
    MereTDD::confirm(true, actual)
#define CONFIRM( expected, actual ) \
    MereTDD::confirm(expected, actual)
```

Once source_location works properly, then we won't really need these macros anymore. The first two are still useful because they eliminate the need to specify the expected bool value. Additionally, all three are slightly useful because they wrap up the specification of the MereTDD namespace. Even though we won't technically need to keep using the macros, I like to keep using them because I think that the all-caps names help the confirmations stand out in the tests better.

This improvement would have been minor and limited to just the confirm functions and macros. So, should we still move to C++ 20 even though we can't yet use source_location? I think so. If nothing else, this bug shows that changes are always being made to the standard libraries, and using the latest compiler and standard library is normally the best choice. Plus, there will be features we will use later in the book that are only found in C++20. For example, we'll be using the std::map class and a useful method that was added in C++20 to determine whether the map contains an element already. We'll be using *concepts* in *Chapter 12, Creating Better Test Confirmations*, which are only found in C++20.

The next improvement will be a bit more involved.

Exploring lambdas for tests

It's getting more and more common for developers to avoid macros in their code. And I agree that there is almost no need for macros anymore. With `std::source_location` from the previous section, one of the last reasons to use macros has been eliminated.

Some companies might even have rules against using macros anywhere in their code. I think that's a bit too much especially given the trouble with `std::source_location`. Macros still have the ability to wrap up code so that it can be inserted instead of the macro itself.

As the previous section shows, the `CONFIRM_TRUE`, `CONFIRM_FALSE`, and `CONFIRM` macros may no longer be absolutely necessary. I still like them. But if you don't want to use them, then you don't have to – at least once `std::source_location` works reliably in a large project.

The `TEST` and `TEST_EX` macros are still needed because they wrap up the declaration of the derived test classes, give them unique names, and set up the code so that the test body can follow. The result looks like we're declaring a simple function. This is the effect we want. A test should be simple to write. What we have now is about as simple as it gets. But the design uses macros. Is there anything we can do to remove the need for the `TEST` and `TEST_EX` macros?

Whatever changes we make, we should keep the simplicity of declaring a test in `Creation.cpp` so that it looks similar to the following:

```
TEST("Test can be created")
{
}
```

What we really need is something that introduces a test, gives it a name, lets the test register itself, and then lets us write the body of the test function. The `TEST` macro provides this ability by hiding the declaration of a global instance of a class derived from the `TestBase` class. This declaration is left unfinished by the macro, so we can provide the body of the test function inside the curly braces. The other `TEST_EX` macro does something similar with the addition of catching the exception provided to the macro.

There is another way to write a function body in C++ without giving the function body a name. And that is to declare a *lambda*. What would a test look like if we stopped using the `TEST` macro and implemented the test function with a lambda instead? For now, let's just focus on tests that do not expect an exception to be thrown. The following is what an empty test might look like:

```
Test test123("Test can be created") = [] ()
{
};
```

With this example, I'm trying to stick to the syntax needed by C++. This assumes we have a class called Test that we want to create an instance of. In this design, tests would reuse the Test class instead of defining a new class. The Test class would override the operator = method to accept a lambda. We need to give the instance a name so that the example uses test123. Why test123? Well, any object instance created still needs a unique name, so I'm using a number to provide something unique. We would need to continue using a macro to generate a unique number based on the line number if we decided to use this design. So, while this design avoids a new derived class for each test, it creates a new lambda for each test instead.

There's a bigger problem with this idea. The code doesn't compile. It might be possible to get the code to compile within a function. But as a declaration of a global Test instance, we can't call an assignment operator. The best I can come up with would be to put the lambda inside the constructor as a new argument, as follows:

```
Test test123("Test can be created", [] ()
{
});
```

While it works for this test, it causes problems in the expected failure tests when we try to call the setExpectedFailureReason method because setExpectedFailureReason is not in scope within the lambda body. Also, we're getting further away from the simple way we have now of declaring a test. The extra lambda syntax and the closing parenthesis and semicolon at the end make this harder to get right.

I've seen at least one other test library that does use lambdas and appears to avoid the need to declare a unique name and, thereby, avoid the need for a macro with something like this:

```
int main ()
{
    Test("Test can be created") = [] ()
    {
    };

    return 0;
};
```

But what this actually does is call a *function* named Test and pass the string literal as an argument. Then, the function returns a temporary object that overrides operator =, which is called to accept the lambda. The only place functions can be called is within other functions or class methods. That means a solution like this needs to declare tests from within a function, and the tests cannot be declared globally as instances like we are doing.

Usually, this means you declare all your tests from within the main function. Or you declare your tests as simple functions and call those functions from within main. Either way, you end up modifying main to call every test function. If you forget to modify main, then your test won't get run. We're going to keep main simple and uncluttered. The only thing main will do in our solution is run the tests that have been registered.

Even though lambdas won't work for us because of the added complexity and because of the inability to call test methods such as setExpectedFailureReason, we can improve the current design a bit. The TEST and, especially, TEST_EX macros are doing work that we can remove from the macros.

Let's start by modifying the TestBase class in Test.h so that it registers itself instead of doing the registration with derived classes in the macros. Also, we need to move the getTests function right before the TestBase class. And we need to forward declare the TestBase class since getTests uses a pointer to TestBase, like this:

```
class TestBase;

inline std::vector<TestBase *> & getTests ()
{
    static std::vector<TestBase *> tests;

    return tests;
}

class TestBase
{
public:
    TestBase (std::string_view name)
    : mName(name), mPassed(true), mConfirmLocation(-1)
    {
        getTests().push_back(this);
    }
```

We'll keep the rest of TestBase unchanged because it handles properties such as the name and whether the test passed or not. We still have derived classes, but the goal of this simplification is to remove any work that the TEST and TEST_EX macros need to perform.

Most of the work that the TEST macro needs to do is to declare a derived class with a run method that will be filled in. The need to register the test is now handled by TestBase. The TEST_EX macro can be simplified further by creating another class called `TestExBase`, which will deal with the expected exception. Declare this new class right after `TestBase`. It looks like this:

```
template <typename ExceptionT>
class TestExBase : public TestBase
{
public:
    TestExBase (std::string_view name,
        std::string_view exceptionName)
    : TestBase(name), mExceptionName(exceptionName)
    { }

    void runEx () override
    {
        try
        {
            run();
        }
        catch (ExceptionT const &)
        {
            return;
        }

        throw MissingException(mExceptionName);
    }

private:
    std::string mExceptionName;
};
```

The `TestExBase` class derives from `TestBase` and is a template class designed to catch the expected exception. This code is currently written into TEST_EX, and we will change TEST_EX to use this new base class instead.

We're ready to simplify the TEST and TEST_EX macros. The new TEST macro looks like this:

```
#define TEST( testName ) \
namespace { \
class MERETDD_CLASS : public MereTDD::TestBase \
```

```
{ \
public: \
    MERETDD_CLASS (std::string_view name) \
    : TestBase(name) \
    { } \
    void run () override; \
}; \
} /* end of unnamed namespace */ \
MERETDD_CLASS MERETDD_INSTANCE(testName); \
void MERETDD_CLASS::run ()
```

It's slightly simpler than before. The constructor no longer needs to have code in the body because the registration is done in the base class.

The bigger simplification is in the TEST_EX macro, which looks like this:

```
#define TEST_EX( testName, exceptionType ) \
namespace { \
class MERETDD_CLASS : public MereTDD::TestExBase<exceptionType> \
\
{ \
public: \
    MERETDD_CLASS (std::string_view name, \
        std::string_view exceptionName) \
    : TestExBase(name, exceptionName) \
    { } \
    void run () override; \
}; \
} /* end of unnamed namespace */ \
MERETDD_CLASS MERETDD_INSTANCE(testName, #exceptionType); \
void MERETDD_CLASS::run ()
```

It's a lot simpler than before because all the exception handling is done in its direct base class. Notice how the macro still needs to use the # operator for exceptionType when constructing the instance. Additionally, notice how it uses exceptionType without the # operator when specifying the template type to derive from.

Summary

This chapter explored ways in which to improve the test library by making use of a new feature in C++ 20 to get line numbers from the standard library instead of from the preprocessor. Even though the new code doesn't work right now, it will eventually make the CONFIRM_TRUE, CONFIRM_FALSE, and CONFIRM macros optional. You will no longer have to use the macros. But I still like to use them because they help wrap up code that is easy to get wrong. And the macros are easier to spot in the tests because they use all capital letters.

We also explored a trend to avoid macros when declaring tests and what it would look like if we used lambdas instead. The approach almost worked with a more complicated test declaration. The extra complexity doesn't matter though because the design did not work for all the tests.

It is still valuable for you to read about the proposed changes. You can learn about how other test libraries might work and understand why this book explains a solution that embraces macros.

This chapter has also shown you how to follow the TDD process at a higher level. The step in the process to enhance a test can be applied to an overall design. We were able to improve and simplify the TEST and TEST_EX macros, which makes all of the tests better.

The next chapter will explore what will be needed to add code that will run before and after the tests to help get things ready for the tests and clean things up after the tests finish.

7
Test Setup and Teardown

Have you ever worked on a project where you needed to prepare your work area first? Once ready, you can finally do some work. Then, after a while, you need to clean up your area. Maybe you use the area for other things and can't just leave your project sitting around or it would get in the way.

Sometimes, tests can be a lot like that. They might not take up table space, but sometimes they can require an environment setup or some other results to be ready before they can run. Maybe a test makes sure that some data can be deleted. It makes sense that the data should exist first. Should the test be responsible for creating the data that it is trying to delete? It would be better to wrap up the data creation inside its own function. But what if you need to test several different ways to delete the data? Should each test create the data? They could call the same setup function.

If multiple tests need to perform similar preparation and cleanup work, not only is it redundant to write the same code into each test, but it also hides the real purpose of the tests.

This chapter is going to allow tests to run preparation and cleanup code so that they can focus on what needs to be tested. The preparation work is called **setup**. And the cleanup is called **teardown**.

We're following a TDD approach, so that means we'll start with some simple tests, get them working, and then enhance them for more functionality.

Initially, we'll let a test run the setup code and then the teardown at the end. Multiple tests can use the same setup and teardown, but the setup and teardown will be run each time for each test.

Once that is working, we'll enhance the design to let a group of tests share setup and teardown code that runs just once before and after the group of tests.

In this chapter, we will cover the following main topics:

- Supporting test setup and teardown
- Enhancing test setup and teardown for multiple tests
- Handling errors in setup and teardown

By the end of the chapter, the tests will be able to have both individual setup and teardown code along with setup and teardown code that encapsulates groups of tests.

Technical requirements

All the code in this chapter uses standard C++, which builds on any modern C++ 20, or later, compiler and standard library. The code is based on and continues from the previous chapters.

You can find all of the code for this chapter at the following GitHub repository:

```
https://github.com/PacktPublishing/Test-Driven-Development-with-CPP
```

Supporting test setup and teardown

In order to support test setup and teardown, we only need to arrange for some code to run before a test begins and for some code to run after a test finishes. For the setup, we might be able to simply call a function near the beginning of the test. The setup doesn't actually have to run before the test, as long as it runs before the test needs the setup results. What I mean is that the unit test library doesn't really need to run the setup before a test begins. As long as the test itself runs the setup at the very beginning of the test, then we get the same overall result. This would be the simplest solution. It's not really a new solution at all, though. A test can already call other functions.

The biggest problem I see with simply declaring a standalone function and calling it at the start of a test is that the intent can get lost. What I mean is that it's up to the test author to make sure that a function called within a test is clearly defined to be a setup function. Because functions can be named anything, unless it has a good name, just calling a function is not enough to identify the intention to have a setup.

What about the teardown? Can this also be a simple function call? Because the teardown code should always be run at the end of a test, the test author would have to make sure that the teardown runs even when an exception is thrown.

For these reasons, the test library should provide some help with the setup and teardown. How much help and what that help will look like is something we need to decide. Our goal is to keep the tests simple and make sure that all the edge cases are handled.

Following a TDD approach, which was first explained in *Chapter 3, The TDD Process*, we should do the following:

- First, think about what the desired solution should be.
- Write some tests that use the solution to make sure it will meet our expectations.
- Build the project and fix the build errors without worrying about getting tests to pass yet.
- Implement a basic solution with passing tests.
- Enhance the solution and improve the tests.

One option to help with setup and teardown would be to add new parameters to the TEST and TEST_EX macros. This would make the setup and teardown part of the test declaration. But is this necessary? If possible, we should avoid relying on these macros. They're already complicated enough without adding more features if we can avoid it. Modifying the macros shouldn't be needed for test setup and teardown.

Another possible solution is to create a method in the TestBase class like we did to set the expected failure reason in *Chapter 3, The TDD Process*. Would this work? To answer that, let's think about what the setup and teardown code should do.

The setup should get things ready for the test. This likely means that the test will need to refer to data or resources such as files that the setup code prepares. It might not seem like much of a setup if the test doesn't get something it can use, but who knows? Maybe the setup does something related but unnoticed by the test code. The main point I'm getting at is that the setup code could do almost anything. It might require its own arguments to customize. Or it might be able to run without any input at all. It might generate something that the test uses directly. Or it might work behind the scenes in a way that is useful but unknown to the test.

Also, the teardown might need to refer to whatever was set up so that it can be undone. Or maybe the teardown just cleans everything up without worrying about where it came from.

Calling a method in TestBase to register and run the setup and teardown seems like it might make the interaction with the test code more complicated because we would need a way to share the setup results. All we really want is to run the setup, gain access to whatever the setup provides, and then run the teardown at the end of the test. There's an easy way to do this that allows whatever interaction is needed between the setup, teardown, and the rest of the test code.

Let's start by creating a new .cpp file in the tests folder called Setup.cpp. The project structure will look like this:

```
MereTDD project root folder
    Test.h
    tests folder
        main.cpp
        Confirm.cpp
        Creation.cpp
        Setup.cpp
```

Here is a test in Setup.cpp that we can use to get started:

```
TEST_EX("Test will run setup and teardown code", int)
{
    int id = createTestEntry();
```

```
        // If this was a project test, it might be called
        // "Updating empty name throws". And the type thrown
        // would not be an int.
        updateTestEntryName(id, "");

        deleteTestEntry(id);
    }
```

The test uses three functions: `createTestEntry`, `updateTestEntryName`, and `deleteTestEntry`. The comment explains what the test might be called and what it would do if this was a test for an actual project instead of a test for the test library. The idea of the test is to call `createTestEntry` to set up some data, try to update the name with an empty string to ensure that is not allowed, and then call `deleteTestEntry` to tear down the data created at the beginning of the test. You can see that the setup provides an identity called `id`, which is needed by the test and by the teardown.

The test expects the call to `updateTestEntryName` to fail because of the empty name. This will result in an exception being thrown. We're just going to throw an int here, but in an actual project, the exception type would normally be something else. The exception will cause the teardown call to `deleteTestEntry` to be skipped.

Additionally, the test could use confirmations to verify its own results if needed. And a failed confirmation will also throw an exception. We need to make sure that the teardown code is run in all cases. Right now, it will always be skipped because the whole purpose of the test is to expect an exception to be thrown from `updateTestEntryName`. But other tests might still skip the teardown if they fail a confirmation.

Even if we fix the problem that `deleteTestEntry` doesn't get called, we still have a test that's unclear. What's really being tested? The only thing in this test that should stand out as the intent of the test is the call to `updateTestEntryName`. The calls to `createTestEntry` and `deleteTestEntry` only hide the real purpose of the test. And if we add a `try`/`catch` block to make sure that `deleteTestEntry` gets called, then the real purpose will only be hidden further.

The three functions in the test are the type of functions that would be found in a project. We don't have a separate project, so they can be placed in `Setup.cpp` because they are helper functions for our purposes. They look like this:

```
#include "../Test.h"

#include <string_view>

int createTestEntry ()
```

```
{
    // If this was real code, it might open a
    // connection to a database, insert a row
    // of data, and return the row identifier.
    return 100;
}


void updateTestEntryName (int /*id*/, std::string_view name)
{
    if (name.empty())
    {
        throw 1;
    }
    // Real code would proceed to update the
    // data with the new name.
}


void deleteTestEntry (int /*id*/)
{
    // Real code would use the id to delete
    // the temporary row of data.
}
```

The id parameter name is commented out because the helper functions don't use them.

We can wrap up the calls to createTestEntry and deleteTestEntry in the constructor and destructor of a class. This helps simplify the test and ensures that the teardown code gets called. The new test looks like this:

```
TEST_EX("Test will run setup and teardown code", int)
{
    TempEntry entry;

    // If this was a project test, it might be called
    // "Updating empty name throws". And the type thrown
    // would not be an int.
    updateTestEntryName(entry.id(), "");
}
```

The TempEntry class contains the setup and teardown calls along with the identifier needed by the test and the teardown. It can go in Setup.cpp right after the three helper methods:

```cpp
class TempEntry
{
public:
    TempEntry ()
    {
        mId = createTestEntry();
    }

    ~TempEntry ()
    {
        deleteTestEntry(mId);
    }

    int id ()
    {
        return mId;
    }

private:
    int mId;
};
```

Writing a class like this is a great way to make sure code gets executed when an instance goes out of scope, and we can use it to make sure that the teardown code always gets run at the end of the test. It's simple and can maintain its own state such as the identifier. Additionally, it only needs a single line to create an instance at the beginning of the test so that it doesn't distract from what the test is trying to do.

You can go this route anytime you have a specific need that the library code doesn't meet. But is there a way that the test library can help make this even better?

I've seen classes that let you pass lambdas or functions to the constructor that do something similar. The constructor will call the first function right away and will call the second when the instance gets destroyed. That's just like what the TempEntry class does except for one detail. TempEntry also manages the identity that is needed by the teardown code. None of the lambda solutions I can think of are as clean as a class written just for this purpose, such as TempEntry. But maybe we can still improve this a little more.

The problem with TempEntry is that it's not clear what is the setup and what is the teardown. It's also not clear in the test that the first line that creates a TempEntry class has anything to do with setup and teardown. Sure, a little studying will let you realize that the setup is in the constructor and the teardown is in the destructor. It would be nice if we had methods called setup and teardown and if the test itself clearly identified the use of the setup and teardown code being run.

One solution that comes to mind would be a base class that calls virtual setup and teardown methods. But we can't use normal inheritance because we need to call them from the constructor and destructor. Instead, we can use a design pattern called *policy-based design*.

A *policy class* implements one or more methods that a derived class will make use of. The methods that the derived class use are called the *policy*. It's like inheritance only backward. We'll turn the TempEntry class into a policy class that implements the setup and teardown methods by modifying it like this:

```
class TempEntry
{
public:
    void setup ()
    {
        mId = createTestEntry();
    }

    void teardown ()
    {
        deleteTestEntry(mId);
    }

    int id ()
    {
        return mId;
    }

private:
    int mId;
};
```

The only real change was to turn the constructor into the setup method and the destructor into the teardown method. That's only because we were using those methods previously to do the work. Now we have a class that is clear and easy to understand. But how will we use it? We no longer have the setup code running automatically when the class is constructed and the teardown code that runs on destruction. We'll need to create another class, and this one can go in `Test.h` because it will be used for all the setup and teardown needs for all the tests. Add this template class inside the `MereTDD` namespace in `Test.h` like this:

```
template <typename T>
class SetupAndTeardown : public T
{
public:
    SetupAndTeardown ()
    {
        T::setup();
    }

    ~SetupAndTeardown ()
    {
        T::teardown();
    }
};
```

The `SetupAndTeardown` class is where we tie the calls to `setup` and `teardown` back into the constructor and destructor. You can use any class you want for the policy as long as that class implements the two `setup` and `teardown` methods. Also, a nice benefit is that because of the public inheritance, you have access to other methods you define in the policy class. We use this to still be able to call the `id` method. A policy-based design lets you extend the interface to whatever you need as long as you implement the policy. In this example, the policy is just the two `setup` and `teardown` methods.

One other thing about using a policy-based design, and specifically about the inheritance, is that this pattern goes against the *is-a* relationship of object-oriented design. If we were using public inheritance in a normal way, then we could say that `SetupAndTeardown` is a `TempEntry` class. In this case, that doesn't make sense. That's okay because we're not going to use this pattern to create instances that can be substituted for one another. We use public inheritance just so that we can call methods such as `id` inside the policy class.

Now that we have all this, what does the test look like? The test can now use the `SetupAndTeardown` class like this:

```
TEST_EX("Test will run setup and teardown code", int)
{
    MereTDD::SetupAndTeardown<TempEntry> entry;

    // If this was a project test, it might be called
```

```
        // "Updating empty name throws". And the type thrown
        // would not be an int.
        updateTestEntryName(entry.id(), "");
}
```

This is a big improvement because of the following list of reasons:

- It's clear at the beginning of the test that we have setup code and teardown code attached to the test
- The teardown code will be run at the end of the test, and we don't need to complicate the test code with a try/catch block
- We don't need to mix calls to `setup` and `teardown` into the rest of the test
- We can interact with the setup results through methods that we write such as the `id` method

Anytime you need setup and/or teardown code within a test, all you have to do is write a class that implements the `setup` and `teardown` methods. If there is no work needed by one of these methods, then leave the method empty. However, both methods need to exist because they are the policy. Implementing the policy methods is what creates a policy class. Then, add an instance of `MereTDD:SetupAndTeardown` that uses the policy class as its template parameter. The test should declare the `SetupAndTeardown` instance at the beginning of the test in order to get the most benefit from this design.

While we can declare the setup and teardown code that runs at the beginning and end of each test like this, we'll need a different solution to share the setup and teardown code so that the setup runs before a group of tests and the teardown code runs after the group of tests complete. The next section will enhance the setup and teardown to meet this expanded need.

Enhancing test setup and teardown for multiple tests

Now that we have the ability to set things up for a test and cleanup after a test, we can do things such as preparing temporary data in the setup that a test needs in order to run and then removing the temporary data in the teardown after a test has run. If there are many different tests using data like this, they can each create similar data.

But what if we need to set up something for a whole group of tests, and then tear it down after all the tests finish? I'm talking about something that remains in place across multiple tests. For the temporary data, maybe we need to prepare a place to hold the data. If the data is stored inside a database, this would be a good time to open the database and make sure the necessary tables are ready to hold the data that each test will be creating. Even the connection to the database itself can remain open and used by the tests. And once all the data tests are done, then the teardown code can close the database.

The scenario applies to many different situations. If you are testing something related to files on a hard drive, then you might want to ensure the proper directories are ready so that the files can be created. The directories can be set up before any of the file tests begin, and the tests only need to worry about creating the files they will be testing.

If you are testing a web service, maybe it makes sense to make sure that your tests have a valid and authenticated login before they begin. It might not make sense for each test to repeat the login steps each time. Unless, of course, that's the purpose of the test.

The main idea here is that while it's good to have some code run as setup and teardown for each test, it can also be good to have different setup and teardown code run only once for a group of tests. That's what this section is going to explore.

We'll call a group of tests that are related by common setup and teardown code that runs once for the entire group a *test suite*. Tests don't have to belong to a test suite, but we'll create an internal and hidden test suite to group all of the individual tests that don't have a specific suite of their own.

We were able to add setup and teardown code to an individual test completely within that test because the setup and teardown code within a test is just like calling a couple of functions. However, in order to support setup and teardown for a test suite, we're going to have to do work outside of the tests. We need to make sure that the setup code runs before any of the related tests run. And then run the teardown code after all of the related tests are complete.

A test project that contains and runs all the tests should be able to support multiple test suites. This means that a test will need some way to identify what test suite it belongs to. Also, we'll need some way to declare the setup and teardown code for a test suite.

The idea works like this: we'll declare and write some code to be a test suite setup. Or maybe we can let the setup code be optional if the only thing needed is the teardown code. Then, we'll declare and write some code for the teardown. The teardown should also be optional. Either the setup or the teardown, or both, need to be defined in order to have a valid test suite. And each test needs some way to identify the test suite it belongs to. When running all the tests in the project, we need to run them in the proper order so that the test suite setup runs first, followed by all the tests in the test suite, and then followed by the test suite teardown.

How will we identify test suites? The test library automatically generates unique names for each test, and these names are hidden from the test author. We could use names for the test suites, too, but let the test author specify what the name should be for each test suite. That seems understandable and should be flexible enough to handle any situation. We'll let the test author provide a simple string name for each test suite.

When dealing with names, one edge case that always comes up is what to do about duplicate names. We'll need to decide. We could either detect duplicate names and stop the tests with an error, or we could stack the setup and teardown so that they all run.

Did we have this duplicate problem with the individual test setup and teardown? Not really because the setup and teardown weren't named. But what happens if a test declares multiple instances of `SetupAndTeardown`? We actually didn't consider that possibility in the previous section. In a test, it might look like this:

```
TEST("Test will run multiple setup and teardown code")
{
    MereTDD::SetupAndTeardown<TempEntry> entry1;
    MereTDD::SetupAndTeardown<TempEntry> entry2;

    // If this was a project test, it might need
    // more than one temporary entry. The TempEntry
    // policy could either create multiple data records
    // or it is easier to just have multiple instances
    // that each create a single data entry.
    updateTestEntryName(entry1.id(), "abc");
    updateTestEntryName(entry2.id(), "def");
}
```

It is an interesting ability to have multiple setup and teardown instances and should help simplify and let you reuse the setup and teardown code. Instead of creating special setup and teardown policy classes that do many things, this will let them be stacked so that they can be more focused. Maybe one test only needs a single piece of data set up and torn down at the end while another test needs two. Instead of creating two different policy classes, this ability will let the first test declare a single `SetupAndTeardown` instance, while the second test reuses the same policy class by declaring two.

Now that we allow individual test setup and teardown code to be composed, why not allow the test suite setup and teardown to be composed, too? This seems reasonable and might even simplify the test library code. How is that possible?

Well, now that we know about the ability, we can plan for it and will likely be able to avoid writing code to detect and throw errors instead. If we notice two or more test suite setup definitions with the same name, we can add them to a collection instead of treating the situation as a special error case.

If we do have multiple setup and teardown definitions with the same name, let's not rely on any particular order between them. They could be defined in different .cpp files just like how the tests can be split between different .cpp files. This will simplify the code because we can add them to a collection as we find them without worrying about a particular order.

The next thing to consider is how to define the test suite setup and teardown code. They probably can't be simple functions because they need to register themselves with the test library. The registration is needed so that when a test gives a suite name, we will know what the name means. The registration seems very similar to how the tests register themselves. We should be able to add an extra string for the suite name. Additionally, the tests will need this new suite name even if they are not part of a specific test suite. We'll use an empty suite name for tests that want to run outside of any test suite.

The registration will need to let tests register themselves with a suite name even if the setup and teardown code for that suite name hasn't yet been registered. This is because tests can be defined in multiple .cpp files, and we have no way of knowing in what order the initialization code will register the tests and test suite setup and teardown code.

There's one more important requirement. We have a way to interact with the setup results in individual test setup and teardown code. We'll need this ability in the test suite setup and teardown, too. Let's say that the test suite setup needs to open a database connection that will be used for all of the tests in the suite. The tests will need some way to know about the connection. Additionally, the test suite teardown will need to know about the connection if it wants to close the connection. Maybe the test suite setup also needs to create a database table. The tests will need the name of that table in order to use it.

Let's create a couple of helper functions in Setup.cpp that will simulate creating and dropping a table. They should look like this:

```cpp
#include <string>
#include <string_view>

std::string createTestTable ()
{
    // If this was real code, it might open a
    // connection to a database, create a temp
    // table with a random name, and return the
    // table name.
    return "test_data_01";
}


void dropTestTable (std::string_view /*name*/)
{
    // Real code would use the name to drop
    // the table.
}
```

Then, in the Setup.cpp file, we can make our first test suite setup and teardown look like this:

```cpp
class TempTable
{
```

```cpp
public:
    void setup ()
    {
        mName = createTestTable();
    }

    void teardown ()
    {
        dropTestTable(mName);
    }

    std::string tableName ()
    {
        return mName;
    }

private:
    std::string mName;
};
```

This looks a lot like the policy class used in the previous section to define the `setup` and `teardown` methods and to provide an interface to access any additional methods or data provided by the setup code. That's because this is going to be a policy class, too. And we might as well make the policies the same no matter if the setup and teardown code is being used for an individual test or an entire test suite.

When we declare that a test has setup and teardown code for just that test, we declare an instance of `MereTDD::SetupAndTeardown` that is specialized with a policy class. This is enough to run the setup code right away and to make sure that the teardown code is run at the end of the test. But in order to gain access to the other information, it's important to give the `SetupAndTeardown` instance a name. The setup and teardown code is fully defined and accessible through the local named instance.

However, with the test suite setup and teardown, we need to put the instances of the policy class into a container. The container will want everything inside it to be of a single type. The setup and teardown instances can no longer be simple local named variables in a test. Yet, we still need a named type because that's how a test in the test suite can access the resources provided by the setup code.

We need to figure out two things. The first is where to create instances of test suite setup and teardown code. And the second is how to reconcile the need for a container to have everything be of a single type with the need for the tests to be able to refer to named instances of specific types that can be different from one policy class to another.

The first problem is the easiest to solve because we need to consider lifetime and accessibility. The test suite setup and teardown instances need to exist and be valid for multiple tests within a test suite. They can't exist as local variables within a single test. They need to be somewhere that they will remain valid for multiple tests. They could be local instances inside of main – that would solve the lifetime issue. But then they would only be accessible to main. The test suite setup and teardown instances will need to be global. Only then will they exist for the duration of the test application and be accessible to multiple tests.

For the second problem, we're going to first declare an interface that the collection will use to hold all the test suite setup and teardown instances. And the test library will also use this same interface when it needs to run the setup and teardown code. The test library needs to treat everything the same because it doesn't know anything about the specific policy classes.

We'll come back to all of this later. Before we get too far, we need to consider what our proposed usage will look like. We are still following a TDD approach, and while it's good to think of all the requirements and what is possible, we're far enough along to have a good idea of what a test suite setup and teardown usage would look like. We even have the policy class ready and defined. Add this to Setup.cpp as the intended usage we're going to implement:

```
MereTDD::TestSuiteSetupAndTeardown<TempTable>
gTable1("Test suite setup/teardown 1", "Suite 1");

MereTDD::TestSuiteSetupAndTeardown<TempTable>
gTable2("Test suite setup/teardown 2", "Suite 1");

TEST_SUITE("Test part 1 of suite", "Suite 1")
{
    // If this was a project test, it could use
    // the table names from gTable1 and gTable2.
    CONFIRM("test_data_01", gTable1.tableName());
    CONFIRM("test_data_01", gTable2.tableName());
}

TEST_SUITE_EX("Test part 2 of suite", "Suite 1", int)
{
    // If this was a project test, it could use
    // the table names from gTable1 and gTable2.
    throw 1;
}
```

There are a few things to explain with the preceding code. You can see that it declares two instances of `MereTDD::TestSuiteSetupAndTeardown`, each specialized with the `TempTable` policy class. These are global variables with a specific type, so the tests will be able to see them and use the methods in the policy class for each. You can use a different policy class for each if you want. Or if you use the same policy class, then there should normally be some difference. Otherwise, why have two instances? For creating temp tables, as this example shows, each table would likely have a unique random name and be able to use the same policy class.

The constructor needs two strings. The first is the name of the setup and teardown code. We're going to treat test suite setup and teardown code as if it was a test itself. We'll include the test suite setup and teardown pass or fail results in the test application summary and identify it with the name provided to the constructor. The second string is the name of the test suite. This can be anything except for an empty string. We'll treat an empty test suite name as a special value for all the tests that do not belong to a test suite.

In this example, both instances of `TestSuiteSetupAndTeardown` use the same suite name. This is okay and supported, as we decided earlier. Anytime there are multiple test suite setup and teardown instances with the same name, they will all be run before the test suite begins.

The reason for a new `TestSuiteSetupAndTeardown` test library class instead of reusing the existing `SetupAndTeardown` class will become clear later. It needs to merge a common interface together with the policy class. The new class also makes it clear that this setup and teardown are for a test suite.

Then come the tests. We need a new macro called `TEST_SUITE` so that the name of the test suite can be specified. Other than the test suite name, the macro will behave almost the same as the existing `TEST` macro. We'll need a new macro for tests belonging to a test suite that also expects an exception. We'll call that one `TEST_SUITE_EX`; it behaves similarly to `TEST_EX` except for the additional test suite name.

There are a lot of changes needed in `Test.h` to support test suites. Most of the changes are related to how the tests are registered and run. We have a base class for tests called `TestBase`, which does the registration by pushing a pointer to `TestBase` to a vector. Because we also need to register test suite setup and teardown code and run tests grouped by their test suite, we'll need to change this. We'll keep `TestBase` as the base class for all tests. But it will now be a base class for test suites, too.

The collection of tests will need to change to a map so that the tests can be accessed by their test suite name. Tests without a test suite will still have a suite name. It will just be empty. Additionally, we need to find the test suite setup and teardown code by the suite name, too. We'll need two collections: one map for the tests and one map for the test suite setup and teardown code. And because we need to refactor the existing registration code out of `TestBase`, we'll create a class called `Test` that will be used for the tests and a class called `TestSuite` that will be used for the test suite setup and teardown code. Both the `Test` and `TestSuite` classes will derive from `TestBase`.

The maps will be accessed with the existing getTests function that will be modified to use a map and a new getTestSuites function. First, include a map at the top of Test.h:

```
#include <map>
#include <ostream>
#include <string_view>
#include <vector>
```

Then, further down, change the part that forward-declares the TestBase class and implements the getTests function to look like this:

```
class Test;
class TestSuite;

inline std::map<std::string, std::vector<Test *>> & getTests ()
{
    static std::map<std::string, std::vector<Test *>> tests;

    return tests;
}

inline std::map<std::string, std::vector<TestSuite *>> &
getTestSuites ()
{
    static std::map<std::string,
            std::vector<TestSuite *>> suites;

    return suites;
}
```

The key to each map will be the test suite name as a string. The value will be a vector of either pointers to Test or pointers to TestSuite. When we register a test or test suite setup and teardown code, we will do so by the test suite name. The first registration for any test suite name will need to set up an empty vector. Once the vector has been set up, the test can be pushed to the end of the vector just like it was done previously. And the test suite setup and teardown code will do the same thing. To make this process easier, we'll create a couple of helper methods that can go in Test.h right after the getTestSuites function:

```
inline void addTest (std::string_view suiteName, Test * test)
{
```

```
    std::string name(suiteName);
    if (not getTests().contains(name))
    {
        getTests().try_emplace(name, std::vector<Test *>());
    }
    getTests()[name].push_back(test);
}


inline void addTestSuite (std::string_view suiteName, TestSuite
* suite)
{
    std::string name(suiteName);
    if (not getTestSuites().contains(name))
    {
        getTestSuites().try_emplace(name,
            std::vector<TestSuite *>());
    }
    getTestSuites()[name].push_back(suite);
}
```

Next is the refactored TestBase class, which has been modified to add a test suite name, stop doing the test registration, remove the expected failure reason, and remove the running code. Now the TestBase class will only hold data that is common between the tests and the test suite setup and teardown code. The class looks like this after the changes:

```
class TestBase
{
public:
    TestBase (std::string_view name, std::string_view
suiteName)
    : mName(name),
      mSuiteName(suiteName),
      mPassed(true),
      mConfirmLocation(-1)
    { }

    virtual ~TestBase () = default;
```

```cpp
        std::string_view name () const
        {
            return mName;
        }

        std::string_view suiteName () const
        {
            return mSuiteName;
        }

        bool passed () const
        {
            return mPassed;
        }

        std::string_view reason () const
        {
            return mReason;
        }

        int confirmLocation () const
        {
            return mConfirmLocation;
        }

        void setFailed (std::string_view reason,
            int confirmLocation = -1)
        {
            mPassed = false;
            mReason = reason;
            mConfirmLocation = confirmLocation;
        }

    private:
        std::string mName;
        std::string mSuiteName;
```

```
        bool mPassed;
        std::string mReason;
        int mConfirmLocation;
    };
```

The functionality that was pulled out of the previous TestBase class goes into a new derived class, called Test, which looks like this:

```
    class Test : public TestBase
    {
    public:
        Test (std::string_view name, std::string_view suiteName)
        : TestBase(name, suiteName)
        {
            addTest(suiteName, this);
        }

        virtual void runEx ()
        {
            run();
        }

        virtual void run () = 0;

        std::string_view expectedReason () const
        {
            return mExpectedReason;
        }

        void setExpectedFailureReason (std::string_view reason)
        {
            mExpectedReason = reason;
        }

    private:
        std::string mExpectedReason;
    };
```

The Test class is shorter because a lot of the basic information now lives in the TestBase class. Also, we used to have a TestExBase class, which needs to be changed slightly. It will now be called TestEx and looks like this:

```cpp
template <typename ExceptionT>
class TestEx : public Test
{
public:
    TestEx (std::string_view name,
        std::string_view suiteName,
        std::string_view exceptionName)
    : Test(name, suiteName), mExceptionName(exceptionName)
    { }

    void runEx () override
    {
        try
        {
            run();
        }
        catch (ExceptionT const &)
        {
            return;
        }

        throw MissingException(mExceptionName);
    }

private:
    std::string mExceptionName;
};
```

The only thing that has really changed with TestEx is the name and the base class name.

Now, we can get to the new TestSuite class. This will be the common interface that will be stored in the map and serve as the common interface for running the setup and teardown code by the test library.

The class looks like this:

```
class TestSuite : public TestBase
{
public:
    TestSuite (
        std::string_view name,
        std::string_view suiteName)
    : TestBase(name, suiteName)
    {
        addTestSuite(suiteName, this);
    }

    virtual void suiteSetup () = 0;

    virtual void suiteTeardown () = 0;
};
```

The TestSuite class doesn't have a runEx method like the Test class. Test suites exist to group tests and to prepare an environment for a test to use, so it doesn't make sense to write setup code that is expected to throw an exception. A test suite doesn't exist to test anything. It exists to prepare for one or more tests that will use the resources that the suiteSetup method gets ready. Also, the teardown code is not intended to test anything. The suiteTeardown code is just supposed to clean up whatever was set up. If there are any exceptions during test suite setup and teardown, we want to know about them.

Additionally, the TestSuite class does not have a run method like the Test class because we need to clearly define what the setup is versus what the teardown is. There is no single block of code to run. There are now two separate blocks of code to run, one at setup time and one at teardown time. So, while a Test class is designed to run something, a TestSuite class is designed to prepare a group of tests with a setup and then clean up after the tests with a teardown.

You can see that the TestSuite constructor registers the test suite setup and teardown code by calling addTestSuite.

We have a function called runTests that currently goes through all the tests and runs them. If we put the code to actually run a single test inside a new function, we can simplify the code that goes through all of the tests and then displays the summary. This will be important because we'll need to run more than tests in the new design. We'll also need to run the test suite setup and teardown code.

Here is a helper function to run a single test:

```cpp
inline void runTest (std::ostream & output, Test * test,
    int & numPassed, int & numFailed, int & numMissedFailed)
{
    output << "------- Test: "
        << test->name()
        << std::endl;

    try
    {
        test->runEx();
    }
    catch (ConfirmException const & ex)
    {
        test->setFailed(ex.reason(), ex.line());
    }
    catch (MissingException const & ex)
    {
        std::string message = "Expected exception type ";
        message += ex.exType();
        message += " was not thrown.";
        test->setFailed(message);
    }
    catch (...)
    {
        test->setFailed("Unexpected exception thrown.");
    }

    if (test->passed())
    {
        if (not test->expectedReason().empty())
        {
            // This test passed but it was supposed
            // to have failed.
            ++numMissedFailed;
            output << "Missed expected failure\n"
```

```cpp
                    << "Test passed but was expected to fail."
                    << std::endl;
        }
        else
        {
            ++numPassed;
            output << "Passed"
                << std::endl;
        }
    }
    else if (not test->expectedReason().empty() &&
        test->expectedReason() == test->reason())
    {
        ++numPassed;
        output << "Expected failure\n"
            << test->reason()
            << std::endl;
    }
    else
    {
        ++numFailed;
        if (test->confirmLocation() != -1)
        {
            output << "Failed confirm on line "
                << test->confirmLocation() << "\n";
        }
        else
        {
            output << "Failed\n";
        }
        output << test->reason()
            << std::endl;
    }
}
```

The preceding code is almost identical to what was in `runTests`. Slight changes have been made to the beginning output that displays the test name. This is to help distinguish a test from the setup and teardown code. The helper function also takes references to the record-keeping counts.

We can create another helper function to run the setup and teardown code. This function will perform almost the same steps for setup and teardown. The main difference is which method to call on the `TestSuite` pointer, either `suiteSetup` or `suiteTeardown`. The helper function looks like this:

```
inline bool runSuite (std::ostream & output,
    bool setup, std::string const & name,
    int & numPassed, int & numFailed)
{
    for (auto & suite: getTestSuites() [name])
    {
        if (setup)
        {
            output << "------- Setup: ";
        }
        else
        {
            output << "------- Teardown: ";
        }
        output << suite->name()
            << std::endl;

        try
        {
            if (setup)
            {
                suite->suiteSetup();
            }
            else
            {
                suite->suiteTeardown();
            }
        }
        catch (ConfirmException const & ex)
        {
```

```
                suite->setFailed(ex.reason(), ex.line());
        }
        catch (...)
        {
            suite->setFailed("Unexpected exception thrown.");
        }

        if (suite->passed())
        {
            ++numPassed;
            output << "Passed"
                << std::endl;
        }
        else
        {
            ++numFailed;
            if (suite->confirmLocation() != -1)
            {
                output << "Failed confirm on line "
                    << suite->confirmLocation() << "\n";
            }
            else
            {
                output << "Failed\n";
            }
            output << suite->reason()
                << std::endl;
            return false;
        }
    }
    return true;
}
```

This function is slightly simpler than the helper function to run a test. That's because we don't need to worry about missed exceptions or expected failures. It does almost the same thing. It tries to run either the setup or teardown, catch exceptions, and update the pass or fail counts.

We can use the two helper functions, `runTest` and `runSuite`, inside the `runTests` function, which will need to be modified like this:

```cpp
inline int runTests (std::ostream & output)
{
    output << "Running "
        << getTests().size()
        << " test suites\n";

    int numPassed = 0;
    int numMissedFailed = 0;
    int numFailed = 0;
    for (auto const & [key, value]: getTests())
    {
        std::string suiteDisplayName = "Suite: ";
        if (key.empty())
        {
            suiteDisplayName += "Single Tests";
        }
        else
        {
            suiteDisplayName += key;
        }
        output << "--------------- "
            << suiteDisplayName
            << std::endl;

        if (not key.empty())
        {
            if (not getTestSuites().contains(key))
            {
                output << "Test suite is not found."
                    << " Exiting test application."
                    << std::endl;
                return ++numFailed;
            }
        }
```

```
            if (not runSuite(output, true, key,
                numPassed, numFailed))
            {
                output << "Test suite setup failed."
                    << " Skipping tests in suite."
                    << std::endl;
                continue;
            }
        }

        for (auto * test: value)
        {
            runTest(output, test,
                numPassed, numFailed, numMissedFailed);
        }

        if (not key.empty())
        {
            if (not runSuite(output, false, key,
                numPassed, numFailed))
            {
                output << "Test suite teardown failed."
                    << std::endl;
            }
        }
    }

output << "---------------------------------\n";
output << "Tests passed: " << numPassed
    << "\nTests failed: " << numFailed;
if (numMissedFailed != 0)
{
    output << "\nTests failures missed: "
        << numMissedFailed;
}
output << std::endl;
```

```
        return numFailed;
    }
```

The initial statement that gets displayed shows how many test suites are being run. Why does the code look at the size of the tests instead of the size of the test suites? Well, that's because the tests include everything, tests with a test suite and those tests without a test suite that get run under a made-up suite called Single Tests.

The primary loop in this function looks at every item in the test map. Previously, these were the pointers to the tests. Now each entry is a test suite name and a vector of test pointers. This lets us go through the tests that are already grouped by the test suite that each test belongs to. An empty test suite name represents the single tests that have no test suite.

If we find a test suite that is not empty, then we need to make sure that there exists at least one entry in the test suite with a matching suite name. If not, then this is an error in the test project and no further tests will be run.

If the test project registers a test with a suite name, then it must also register the setup and teardown code for that suite. Assuming we have the setup and teardown code for the suite, each registered setup is run and checked for an error. If there is an error setting up the test suite, then only the tests in that suite will be skipped.

Once all the test suite setup code is run, then the tests are run for that suite.

And after all the tests for the suite are run, then all the test suite teardown code is run.

There are two more parts to enable all this. The first is the TestSuiteSetupAndTeardown class, which goes into Test.h right after the existing SetupAndTeardown class. It looks like this:

```
template <typename T>
class TestSuiteSetupAndTeardown :
    public T,
    public TestSuite
{
public:
    TestSuiteSetupAndTeardown (
        std::string_view name,
        std::string_view suite)
    : TestSuite(name, suite)
    { }

    void suiteSetup () override
```

```
    {
        T::setup();
    }

    void suiteTeardown () override
    {
        T::teardown();
    }
};
```

This is the class that is used in a test .cpp file to declare a test suite setup and teardown instance with a specific policy class. This class uses multiple inheritances to bridge the policy class and the common TestSuite interface class. When the runSuite function calls either suiteSetup or suiteTeardown through a pointer to TestSuite, these virtual methods will end up calling the override methods in this class. Each one just calls the setup or teardown methods in the policy class to do the actual work.

The last change to explain is the macros. We need two additional macros to declare a test that belongs to a test suite without an expected exception and with an expected exception. The macros are called TEST_SUITE and TEST_SUITE_EX. There are minor changes needed in the existing TEST and TEST_EX macros because of the refactoring of the TestBase class. The existing macros need to be updated to use the new Test and TestEx classes instead of TestBase and TestExBase. Additionally, the existing macros need to now pass an empty string for the test suite name. I'll show the new macros here because they are so similar except for the difference in the test suite name. The TEST_SUITE macro looks like this:

```
#define TEST_SUITE( testName, suiteName ) \
namespace { \
class MERETDD_CLASS : public MereTDD::Test \
{ \
public: \
    MERETDD_CLASS (std::string_view name, \
      std::string_view suite) \
    : Test(name, suite) \
    { } \
    void run () override; \
}; \
} /* end of unnamed namespace */ \
MERETDD_CLASS MERETDD_INSTANCE(testName, suiteName); \
void MERETDD_CLASS::run ()
```

The macro now accepts a `suiteName` parameter that gets passed to the instance as the suite name. And the `TEST_SUITE_EX` macro looks like this:

```
#define TEST_SUITE_EX( testName, suiteName, exceptionType ) \
namespace { \
class MERETDD_CLASS : public MereTDD::TestEx<exceptionType> \
{ \
public: \
    MERETDD_CLASS (std::string_view name, \
        std::string_view suite, \
        std::string_view exceptionName) \
    : TestEx(name, suite, exceptionName) \
    { } \
    void run () override; \
}; \
} /* end of unnamed namespace */ \
MERETDD_CLASS MERETDD_INSTANCE(testName, suiteName,
#exceptionType); \
void MERETDD_CLASS::run ()
```

The new suite macros are so similar to the modified non-suite macros that I tried to change the non-suite macros to call the suite macros with an empty suite name. But I could not figure out how to pass an empty string to another macro. The macros are short, so I left them with similar code.

That's all of the changes needed to enable the test suites. The summary output looks a little different now after these changes. Building and running the test project produces the following output. It's a bit long because we have 30 tests now. So, I'm not showing the entire output. The first part looks like this:

```
Running 2 test suites
--------------- Suite: Single Tests
------- Test: Test will run setup and teardown code
Passed
------- Test: Test will run multiple setup and teardown code
Passed
------- Test: Test can be created
Passed
------- Test: Test that throws unexpectedly can be created
Expected failure
Unexpected exception thrown.
```

Here, you can see that there are two test suites. One is named `Suite 1` and contains two tests plus the suite setup and teardown, and the other is unnamed and contains all the other tests that do not belong to a test suite. The first part of the output happens to be the single tests. The rest of the summary output shows the test suite and looks like this:

```
--------------- Suite: Suite 1
------- Setup: Test suite setup/teardown 1
Passed
------- Setup: Test suite setup/teardown 2
Passed
------- Test: Test part 1 of suite
Passed
------- Test: Test part 2 of suite
Passed
------- Teardown: Test suite setup/teardown 1
Passed
------- Teardown: Test suite setup/teardown 2
Passed

----------------------------------
Tests passed: 30
Tests failed: 0
Tests failures missed: 1
```

Each test suite is introduced in the summary output with the suite name followed by all the tests in that suite. For an actual suite, you can see the setup and teardown that surround all the tests. Each setup and teardown is run as if it was a test.

And the end shows the pass and fail counts just like before.

In this section, I briefly explained some of the error handling for the setup and teardown code, but it needs more. The main purpose of this section was to get the setup and teardown working for the test suites. Part of that required a bit of error handling such as what to do when a test declares that it belongs to a suite that doesn't exist. The next section will go deeper into this.

Handling errors in setup and teardown

Bugs can be found anywhere in code, and that includes inside the setup and teardown code. So, how should these bugs be handled? In this section, you'll see that there is no single way to deal with bugs in setup and teardown. It's more important that you be aware of the consequences so that you can write better tests.

Let's start at the beginning. We've already sidestepped a whole class of problems related to multiple setup and teardown declarations. We decided that these would simply be allowed instead of trying to prevent them. So, a test can have as many setup and teardown declarations as it wants. Additionally, a test suite can declare as many instances of setup and teardown as needed.

However, just because multiple instances are allowed doesn't mean that there won't be any problems. The code that creates the test data entries is a good example. I thought about fixing the problem in the code but left it so that I can explain the problem here:

```
int createTestEntry ()
{
    // If this was real code, it might open a
    // connection to a database, insert a row
    // of data, and return the row identifier.
    return 100;
}
```

The problem is hinted at in the preceding comment. It says that real code would return the row identifier. Since this is a test helper function that has no connection to an actual database, it simply returns a constant value of 100.

You'll want to avoid your setup code doing anything that can conflict with additional setup code. A row identity in a database will not conflict because the database will return a different ID each time data is inserted. But what about other fields that get populated in the data? You might have constraints in your table, for example, where a name must be unique. If you create a fixed test name in one setup, then you won't be able to use the same name in another.

Even if you have different fixed names in different setup blocks so they won't cause conflicts, you can still run into problems if the test data doesn't get cleaned up properly. You might find that your tests run okay the first time and then fail thereafter because the fixed names already exist in the database.

I recommend that you randomize your test data. Here is the other example that creates a test table:

```
std::string createTestTable ()
{
    // If this was real code, it might open a
    // connection to a database, create a temp
    // table with a random name, and return the
    // table name.
    return "test_data_01";
}
```

The comment in the preceding code also mentions creating a random name. It's fine to use a fixed prefix, but consider making the digits at the end random instead of fixed. This won't completely solve the problem of colliding data. It's always possible that random numbers will turn out to be the same. But doing this together with a good cleanup of the test data should help eliminate most cases of conflicting setups.

Another problem has already been handled in the test library code. And that is what to do when a test declares that it belongs to a test suite and that test suite does not have any setup and teardown code defined.

This is treated as a fatal error in the test application itself. The moment a required test suite setup and teardown registration cannot be found, the test application exits and does not run any more tests.

The fix is simple. Make sure you always define test suite setup and teardown code for all test suites that the tests use. It's okay to have a test suite setup and teardown code registered that is never used by any test. But the moment a test declares that it belongs to a test suite, then that suite becomes required.

Now, let's talk about exceptions in setup and teardown code. This includes confirmations because a failed CONFIRM macro results in an exception being thrown. It's okay to add confirmations to set up code like this:

```
class TempEntry
{
public:
    void setup ()
    {
        mId = createTestEntry();
        CONFIRM(10, mId);
    }
}
```

Currently, this will cause the setup to fail because the identity is fixed to always be a value of 100. And the confirmation tries to make sure that the value is 10. Because the test setup code is called as if it was a regular function call, the result of this failed confirmation will be the same as any other failed confirmation in the test itself. The test will fail, and the summary will show where and why the failure happened. The summary looks like this:

```
------- Test: Test will run multiple setup and teardown code
Failed confirm on line 51
    Expected: 10
    Actual   : 100
```

However, putting confirmations in the teardown code is not recommended. And throwing exceptions from the teardown code is not recommended – especially for test teardown code because test teardown code is run from inside a destructor. So, moving the confirmation to the teardown code like this will not work the same way:

```
class TempEntry
{
public:
    void setup ()
    {
        mId = createTestEntry();
    }

    void teardown ()
    {
        deleteTestEntry(mId);
        CONFIRM(10, mId);
    }
}
```

This will result in an exception being thrown during the destruction of the `SetupAndTeardown` class that uses the `TempEntry` policy class. The entire test application will be terminated like this:

```
Running 2 test suites
--------------- Suite: Single Tests
------- Test: Test will run setup and teardown code
terminate called after throwing an instance of
'MereTDD::ActualConfirmException'
/tmp/codelite-exec.sh: line 3: 38155 Abort trap:
6              ${command}
```

The problem is not as severe in the test suite teardown code because that teardown code is run by the test library after all the tests in the suite have been completed. It is not run as part of a class destructor. It's still good advice to follow about not throwing any exceptions in the teardown code at all.

Treat your teardown code as an opportunity to clean up the mess left behind by the setup and the tests. Normally, it should not contain anything that needs to be tested.

The test suite setup code is a little different from the test setup code. While an exception in the test setup code causes the test to stop running and fail, an exception thrown in a test suite setup will cause all the tests in that suite to be skipped. Adding this confirmation to the test suite setup will trigger an exception:

```
class TempTable
{
public:
    void setup ()
    {
        mName = createTestTable();
        CONFIRM("test_data_02", mName);
    }
```

And the output summary shows that the entire test suite has been disrupted like this:

```
--------------- Suite: Suite 1
------- Setup: Test suite setup/teardown 1
Failed confirm on line 73
    Expected: test_data_02
    Actual  : test_data_01
Test suite setup failed. Skipping tests in suite.
```

The preceding message says that the test suite will be skipped.

All the error handling that went into the test library for both test setup and teardown and test suite setup and teardown is largely untested itself. What I mean is that we added an extra feature to the test library to support any expected failures. And I did not do the same thing for expected failures in the setup and teardown code. I felt that the extra complexity needed to handle expected failures in the setup and teardown code was not worth the benefit.

We're using TDD to help guide the design of the software and to improve the quality of the software. But TDD doesn't completely remove the need for some manual testing of edge conditions that are too difficult to test in an automated manner or that are just not feasible to test.

So, will there be a test to make sure that the test library really does terminate when a required test suite is not registered? No. That seems like the kind of test that is best handled through manual testing. There might be situations you'll encounter that are similar, and you'll have to decide how much effort is needed to write a test and whether the effort is worth the cost.

Summary

This chapter completes the minimum functionality needed in a unit testing library. We're not done developing the testing library, but it now has enough features to be useful to another project.

You learned about the issues involved with adding setup and teardown code and the benefits provided. The primary benefit is that tests can now focus on what is important to be tested. Tests are easier to understand when everything needed to run the test is no longer cluttering the test and the cleanup is handled automatically.

There are two types of setup and teardown. One is local to the test; it can be reused in other tests but local means that the setup runs at the beginning of the test and the teardown happens at the end of the test. Another test that shares the same setup and teardown will repeat the setup and teardown in that other test.

The other type of setup and teardown is actually shared by multiple tests. This is the test suite setup and teardown; its setup runs before any of the tests in the suite begin, and its teardown runs after all the tests in the suite are complete.

For the local tests, we were able to integrate them into the tests fairly easily without much impact on the test library. We used a policy-based design to make writing the setup and teardown code easy. And the design lets the test code access the resources prepared in the setup.

The test suite setup and teardown was more elaborate and needed extensive support from the test library. We had to change the way tests were registered and run. But at the same time, we simplified the code and made it better. The test suite setup and teardown design uses the same policy that the local setup and teardown uses, which makes the whole design consistent.

And you also learned a few tips about how to handle errors in the setup and teardown code.

The next chapter will continue to give you guidance and tips on how to write better tests.

8

What Makes a Good Test?

A project developed with TDD will have a lot of tests. But don't assume that more or longer tests are always better. You need to have good tests. But what makes a good test?

We're not going to be writing more code in this chapter. This chapter is more of a look back at some of the situations we've already encountered as well as referring to some tests in upcoming chapters. This is a chance to reflect on what you've learned so far and to look forward to upcoming topics.

A good test should incorporate the following elements:

- Be easy to understand – a good understanding will lead to better ideas for more tests and make tests easier to maintain.

- Be focused on a specific scenario – don't try to test everything in one giant test. Doing too much in a test will break the first guidance of understandability.

- Be repeatable – tests that use random behavior to sometimes catch problems can miss issues at the worst times.

- Be kept close to the project – make sure that tests belong to the project they are testing.

- Should test what should happen instead of how it happens – if a test relies too much on internal workings, then it will be brittle and cause more work when the code is refactored.

Each of these topics will be explained in this chapter with examples.

Technical requirements

All code in this chapter comes from other chapters in this book and is used here as example code to reinforce ideas that make good tests.

Making tests easy to understand

Using descriptive test names is probably the single best thing you can do to improve your tests. I like to name my tests using simple sentences whenever possible. For example, one of the earliest tests we created is called `"Test will pass without any confirms"` and looks like this:

```
TEST("Test will pass without any confirms")
{
}
```

A good pattern to follow is this:

```
<object><does something><qualification>
```

Each of the three sections should be replaced with something specific to what you're doing. For the example just given, `<object>` is Test, `<does something>` is `will pass`, and `<qualification>` is `without any confirms`.

I don't always follow this pattern, especially when testing an object or a type for several different and related results. For example, a simple test immediately following the previous test looks like this:

```
TEST("Test bool confirms")
{
    bool result = isNegative(0);
    CONFIRM_FALSE(result);

    result = isNegative(-1);
    CONFIRM_TRUE(result);
}
```

For this simple test, there are only two possibilities. The bool is either false or true. The test is focused on the bool type, and the name fully describes what the test does. My advice is to follow the naming pattern when it makes sense.

The following are ways in which descriptive names help improve your tests:

- The name is what you see in the summary description and will help anyone understand what is being tested with just a glance.

- A descriptive name will help you spot holes in your testing because it's easier to see what's missing when you can clearly see what is already being tested.

- A descriptive name that follows the pattern given will help you focus when writing the test. It's easy to lose track of what a test is supposed to do and start including other things. A descriptive name will help you put related checks in their own tests where they will no longer clutter what is being tested and will have their own descriptive name, which will help them stand out.

Putting all three benefits together, you get a feedback loop that reinforces the need for good naming. You'll naturally create more tests because each one is focused. This leads to a better understanding of what is being tested, which helps you to find missing tests. And then, when writing the new tests, you'll stay on track and create yet more tests as new ideas come to mind.

Imagine if, instead, we had taken a different approach and created a test called `"Confirm"`. What would it do? Does it inspire you to think of more tests? And what code would you write in the test? This is a name that stops the cycle of better tests. No one will know what the test does without reading the code. No one will think of new scenarios because the focus is dragged into the code. And the test code itself will likely be all over the place yet still not cover everything that should be tested.

And let's not forget that using TDD is supposed to help drive our designs, improve the quality of our software, and let us refactor and enhance the code with confidence. Descriptive names help with all of this.

You might find that after a major refactoring, certain tests are no longer applicable. That's fine, and they can be deleted. Descriptive names will help us spot these outdated tests. Sometimes, instead of deleting tests, they can be updated, and tests that are focused will be easier to update.

The next step to creating good tests is to keep them simple. A complicated test is usually a symptom of a bad design. TDD helps improve designs. So when you find a complicated test, that's your signal to simplify the design of the project being tested.

If you can make changes that simplify the tests, then that's usually a double win. You get tests that are easier to understand, which leads to higher quality software that is easier to use. Remember that a test is a consumer of your software, just like any other component. So when you can simplify a test, you're also simplifying other code that uses the same code being tested.

A big part of simplifying tests is to make use of setup and teardown code. This lets the test focus on what it needs to do and lets us read and understand the main point of the test without getting distracted with other code that just gets things ready.

For example, in *Chapter 14, How To Test Services*, I show you the test that I first created to test a service. The test created a local instance of the service and called `start`. I realized that other tests would likely need to start a service, so they might as well share the same service that has already been started with some setup code. The new test uses a test suite that allows multiple tests to share the same setup and teardown code. The test looks like this:

```
TEST_SUITE("Request can be sent and response received",
  "Service 1")
```

```
{
    std::string user = "123";
    std::string path = "";
    std::string request = "Hello";
    std::string expectedResponse = "Hi, " + user;

    std::string response = gService1.service().handleRequest(
        user, path, request);
    CONFIRM_THAT(response, Equals(expectedResponse));
}
```

This test has a descriptive name and focuses on what needs to be tested instead of what is needed to create and start the service. The test uses the global gService1 instance, which exposes the already running service through the service method.

By providing descriptive names and keeping your tests as simple as possible, you'll find better results with TDD that will lead to better software designs. The next section goes into more detail about how to focus on specific scenarios.

Keeping tests focused on specific scenarios

The previous section explained that one of the benefits of descriptive names is that they help keep your tests focused. We're going to explore in this section various scenarios that will give you ideas for what to focus on.

Saying that a test should be focused is great. But if you don't know how to figure out what to focus on, then it won't help you. The advice becomes empty and frustrating.

These five cases will make the advice more meaningful. Not all of them may apply to all situations. But having these will help you, sort of like a checklist. All you need to do is think about each one and write specific tests that cover the case. The cases are as follows:

1. **Happy or normal**: This is a common use case.
2. **Edge**: This is a case near the transition between normal and error cases.
3. **Error**: This is a common problem that needs to be handled.
4. **Not normal**: This is a valid but uncommon use case.
5. **Deliberate misuse**: This is an error case designed to cause problems on purpose.

Let's start with the happy or normal case first. This one should be easy, but it often gets over-complicated by including some of the other cases in the same test. Or another way it can be over-complicated is by creating a test that is too vague or not clear that it's the happy or normal case.

The actual name for this should probably be the normal case since that matches the style of the other cases. But I so often think of this as the happy case that I included both names. You might also think of this as a typical case. However you think of it, all you need to do is pick a scenario that best describes a common way that your code will be used. I think of it as a happy case because there should not be any errors. This should represent usage that is expected and typical and should succeed. For example, in *Chapter 13, How to Test Floating Point and Custom Values*, there's a test for float values that covers 1,000 typical values from 0.1 up to 100 in increments of 0.1. The test looks like this:

```
TEST("Test many float comparisons")
{
    int totalCount {1'000};
    int passCount = performComparisons<float>(totalCount);
    CONFIRM_THAT(passCount, Equals(totalCount));
}
```

An edge case is right on the borderline between a happy case and a problem or error case. You might often need two edge cases where one represents the most extreme usage that is still within normal bounds, and the other is the beginning of the error conditions. An edge case is a transition between good and bad results. And there can often be multiple edge cases.

Edge cases are extremely valuable to include in your testing because they tend to find a lot of bugs, and maybe even more importantly, they make you think about your design. When you consider edge cases, you'll often either accept the edge case or change your design so that the edge case doesn't apply anymore.

The edge cases for the previous float comparisons are to test a very small float value and a very large float value. These are two separate tests and look like this:

```
TEST("Test small float values")
{
    // Based on float epsilon = 1.1920928955078125e-07
    CONFIRM_THAT(0.000001f, NotEquals(0.000002f));
}

TEST("Test large float values")
{
    // Based on float epsilon = 1.1920928955078125e-07
    CONFIRM_THAT(9'999.0f, Equals(9'999.001f));
}
```

Edge cases can sometimes be a bit more technical because there's usually a reason for the test to be an edge case. For float values, the edge cases are based on the *epsilon* value. Epsilon values are explained in *Chapter 13, How to Test Floating Point Values*. Adding tests for small and large floating point values will cause us to change the entire way that we compare floating point values in *Chapter 13*. This is why edge cases are so valuable in testing.

The error case is like the happy case turned sad. Think of a typical problem that your code might need to handle and write a test for that specific problem. Just like how the happy case can sometimes be over-complicated, this one, too, can be over-complicated. You don't need to include minor variations of an error case just for the sake of variation alone. Just pick what you think represents the most common or middle case that should result in an error and create a test for just that one case. Of course, you will want to name the test with a descriptive name that explains the case.

For example, in *Chapter 11, Managing Dependencies*, there's a normal test to make sure that tags *can* be used to filter messages. An error case is almost the opposite and makes sure that an overridden default tag is *not* used to filter the message. The test might not make sense without first reading *Chapter 11*. I'm including it here as an example of an error case. Notice CONFIRM_FALSE at the end of the test, which is the part that ensures the log message does not appear in the log file. The test looks like this:

```
TEST("Overridden default tag not used to filter messages")
{
    MereTDD::SetupAndTeardown<TempFilterClause> filter;
    MereMemo::addFilterLiteral(filter.id(), info);

    std::string message = "message ";
    message += Util::randomString();
    MereMemo::log(debug) << message;

    bool result = Util::isTextInFile(message,
        "application.log");
    CONFIRM_FALSE(result);
}
```

If there are multiple error cases that you think are all important enough to be included, put them in separate tests and ask what makes each one different. This will lead to the insight that might cause you to change your design or might lead to more tests.

I like to include a few tests that are just outside of a normal case but not borderline or edge. These are still within valid use that should succeed but might cause your code to do a little extra work. This case can be valuable in helping to catch regressions. A regression is a bug that is new and represents a problem that used to work previously. A regression is most common after making a large design change. Having some tests that are not normal but still expected to succeed will improve your confidence in your code continuing to work after major changes are made.

The last case is deliberate misuse and is important for security reasons. This is not just an error case; it's an error case crafted to try to cause your code to fail in predictable ways that an attacker can use for their own purposes. For cases like this, instead of creating tests that you know will fail, try to think of what would cause your code to fail spectacularly. Maybe your code treats negative numbers as errors. Then for deliberate misuse, maybe consider using really large negative numbers.

In *Chapter 14, How to Test Services*, there is mention of a possible test for deliberate misuse. We don't actually create this test, but I do describe what the test would look like. In the service, there is a string value that represents the request being made. The code handles unrecognized request strings, and I mentioned that a good test would try to call the service with some request that doesn't exist to make sure that the service properly handles the ill-formed request.

For a final piece of advice about focusing on specific scenarios, I'd like to recommend that you avoid retesting. This is not one of the five cases just mentioned because it applies to all of them.

Retesting is where you check the same thing or make the same confirmations over and over in many tests. You don't need to do this, and it just distracts you from what each test should be focused on.

If there's a property or result that you need to confirm works, then create a test for it. And then you can trust that it will be tested. You don't need to confirm again and again that the property works as expected each time it's used. Once you have a test for something, then you don't need to verify it works in other tests.

Use random behavior only in this way

The previous chapter mentioned using random behavior in tests, and it's important for you to understand more about this so that your tests are predictable and repeatable.

Predictability and randomness are about as opposite as you can get. How should we reconcile these two properties? The first thing to understand is that the tests you write should be predictable. If a test passes, then it should always pass unless something outside of your control fails, such as a hard drive crashing in the middle of your tests. There's no way to predictably handle accidents like that, and that's not what I'm talking about. I mean that if a test passes, then it should continue to pass until some code change causes it to fail.

And when a test fails, then it should continue to fail until the problem is fixed. The last thing you want is to add random behavior to your tests so that you sometimes do one thing and other times do another. That's because the first behavior might pass, while the second behavior goes down a different code path that fails.

If you get a failure, make a change that you think fixes the problem, and then get a pass, you might think that your change fixed the problem. But what if the second test run just happened to use the random behavior that was always passing? It makes it hard to verify that a code change actually fixed a problem.

And worse yet, what happens when some random failure condition never gets run? You think all possible code path combinations are being run, but by chance, one or more conditions are skipped. This can cause you to miss a bug that should have been caught.

I hope I've convinced you to stay away from random test behavior. If you want to test different scenarios, then write multiple tests so that each scenario is covered by its own test that will reliably be run.

Why, then, did I mention using randomness in the previous chapter? I actually do suggest that you use randomness but not in a way that determines what a test does; rather, in a way that helps prevent collisions between different test runs. The random behavior is mentioned in one of the helper functions to create a temporary table like this:

```
std::string createTestTable ()
{
    // If this was real code, it might open a
    // connection to a database, create a temp
    // table with a random name, and return the
    // table name.
    return "test_data_01";
}
```

Let's say that you have a test that needs some data. You create the data in setup and delete it when the test is done in teardown. What happens if the test program crashes during the test and the teardown never gets a chance to delete the data? The next time you run your test, it's likely that the setup will fail because the data still exists. This is the collision I'm talking about.

Maybe you think that you can just enhance the setup to succeed if it finds the data already exists. Well, you can also get collisions in other ways, such as when writing code in a team and two team members happen to run the tests at almost the same time. Both setup steps run, and one of them finds the data already exists and continues. But before the test can begin to use the data, the other team member finished, and the teardown code deletes the data. The team member that is still running tests will now fail because the data no longer exists.

You can almost entirely eliminate this problem by generating random data. Not so random that the behavior of the test is affected. But just random enough to avoid conflicts. Maybe the data is identified by a name. As long as the name is not part of what is being tested, the name can be changed slightly so that each time the test is run, the data will have a different name. The createTestTable function returns a hardcoded name, but the comment mentions that a random name might be better.

There is a place for using full random behavior in tests, such as when performing random penetration testing, and you need to fuzz or alter the data to simulate scenarios that you otherwise would not be able to write specific test cases for. The number of possible combinations could be too many to handle with specific named test cases. So in these situations, it is a good idea to write tests that use random data that can change the behavior and outcome of the tests. But these types of tests won't help you with TDD to improve your designs. They have a place that supplements TDD.

When writing tests that use random behavior, such as when handling uncountable combinations, you'll need to capture the failures because each one will need to be analyzed to find the problem. This is a time-consuming process. It's valuable, but not what you need when writing a test to help you figure out what design to use or when evaluating the results of a major design change to see if anything was broken.

For the types of tests that are most beneficial to TDD, avoid any random behavior that can change the outcome of the tests. This will keep your tests reproducible and predictable.

Only test your project

Other components and libraries will be used and might fail. How should you handle these failures in your tests? My advice is to assume that only your code needs to be tested. You should assume that the components and libraries you are using have already been tested and are working correctly.

For one thing, remember that we are using TDD to improve our own code. If you were to write a test for some code that you bought or found online, how would this affect your own code?

There's always a possibility that you are using an open source library and you have a good idea for improvement. That's great! But that improvement belongs in that other project. It has no place in your own project's tests. Even if you find a bug in a commercial software package, all you can do is report the problem and hope it gets fixed.

The last thing you want to do is put a confirmation in your own test project that confirms some other code is working as expected. This not only does not affect your own designs, it actually makes your tests less focused. It takes away from the clarity you should be aiming for by adding distractions that don't directly benefit your own project.

The next chapter of this book begins *Part 2*, where we'll build a logging library. The logging library will be a separate project with its own set of tests. The logging library will also use the testing library we've been building. Imagine how confusing it would be if we were to add a new feature to the testing library and then test that new feature from the logging library.

Test what should happen instead of how

A common problem I see is when a test tries to verify expected results by checking internal steps along the way. The test is checking how something is done. This type of test is fragile and often needs frequent updates and changes.

A better approach is to test what happens as a final result. Because then the internal steps can change and adapt as needed. The test remains valid the entire time without further maintenance.

If you find yourself going through your tests and frequently updating them, so the tests pass again, then your tests might be testing how something is done instead of what is done.

For example, in *Chapter 10, The TDD Process in Depth*, there's a section called *When is testing too much?* that explains the idea of what to test in greater detail.

The general idea is this. Let's say you have a function that adds a filter to a collection. If you write a test that's focused on how the code works, then you might go through the items in the collection to make sure the filter that was just added really appears in the collection. The problem with this approach is that the collection is an internal step and might change, which will cause the test to fail. A better approach is to first add the filter and then try to perform an operation that would be affected by the filter. Make sure that the filter affects the code as you expect, and leave the internal details of how it works to the code being tested. This is testing what should happen and is a much better approach.

Summary

This has been more of a reflective chapter where you learned some tips to help you write better tests. Examples from earlier and later chapters were used to help reinforce the ideas and guidance. You'll write better tests if you make sure to consider the following items:

- The tests should be easy to understand with descriptive names.
- Prefer small and focused tests instead of large tests that try to do everything.
- Make sure that tests are repeatable. If a test fails once, then it should continue to fail until the code is fixed.
- Once you test something, you don't need to keep testing the same thing. And if you have some useful code that other tests can share, then consider putting the code in its own project with its own set of tests. Only test the code that is in your project.
- Test what should happen instead of how it should happen. In other words, focus less on the internal steps and instead verify the results you are most interested in.

There are many ways to write better tests. This chapter should not be considered to include the only things you need to consider. Instead, this chapter identifies some common issues that cause problems with tests and gives you tips and advice to improve your tests. In the next chapter, we're going to use TDD to create something that will use a unit test library just like any other project you'll be working on.

Part 2:
Using TDD to Create
a Logging Library

This book is divided into three parts. In this second part, we're going to use the unit test library to design and build a logging library. Along the way, you'll see how to use TDD in a different project just as you'll need to do when working on your own projects.

The following chapters are covered in this part:

- *Chapter 9, Using Tests*
- *Chapter 10, The TDD Process in Depth*
- *Chapter 11, Managing Dependencies*

9

Using Tests

Everything in this book until now has been about using TDD to design and build a unit test library. While that has been valuable, it's been a bootstrap project where we've used TDD to help create a tool for TDD.

This chapter is different. For the first time, we're going to use TDD to create something that will use a unit test library just like any other project you'll be working on. This will still be a library that's intended to be used by other projects. We're going to create a logging library, and we're going to use TDD and the unit test library to reach the following goals:

- Create a design that's easy to use
- Improve the quality of the code
- Refactor the design as needed while maintaining confidence in the code
- Create tests that help capture the requirements and document the usage of the library

The approach should be familiar by now. We'll start out simple and get something working, before enhancing the design and adding new features. Each step will start with tests that will drive the design so that we can reach the goals.

We'll first think about why we are building a logging library. This is important to set the overall direction of the project. Then we'll explore how TDD will help us and what the ideal logging library would look like. And then, we'll start building the logging library.

At the end of this chapter, you'll have a simple logging library that you can use in your projects. But even more important is the skill you'll gain by seeing exactly how to use TDD in a real project.

Technical requirements

All code in this chapter uses standard C++ that builds on any modern C++ 20 or later compiler and Standard Library. The code uses the testing library from the previous chapters and will start a new project.

You can find all the code for this chapter in the following GitHub repository:

`https://github.com/PacktPublishing/Test-Driven-Development-with-CPP`

Why build a logging library?

There are already lots of choices available in other libraries for logging. So why build another logging library? Isn't this supposed to be a book about TDD?

This is a practical book that shows you how to use TDD in your own projects. And one of the best ways to do that is to use TDD to build a project. We needed a project for this book, and I thought that a logging library would be perfect because we can start out simple and enhance it along the way. A logging library is useful and practical in itself, which also fits the theme of this book.

If you already have your own logging library or are using one you found elsewhere, then this book will still help you to better understand how it works. And you'll also benefit from the process of building a logging library so that you can apply the same process to your own project.

But we're not going to settle for a logging library that does what all the other logging libraries do. Following a TDD approach, we need to have a good idea of how something will be used. We approach problems from a user's point of view.

We're going to approach this as if we were building micro-services. Instead of building a software solution as one giant application that does everything needed, a micro-service architecture builds smaller applications that accept requests, perform some desired service, and return the results. One service can sometimes call other services to perform its function.

There are many benefits to an architecture such as this that this book will not go into. This architecture is being used to give us a specific use in mind as we employ TDD to design a logging library that will be uniquely suited to a micro-service environment.

Without a focus on and knowledge of the specific user that will need the software you are designing, you face the risk of designing something that doesn't meet the needs of that user. Our user for this book will be a software developer who is writing micro-services and needs to log what the services do. That's the type of focus I encourage you to get to when designing your own software. Because without the focus just described, we might have jumped right into a logging library that does what most other logging libraries do. Our micro-service developer would look at a general-purpose logging library and find nothing special about it.

Our goal instead is for that same micro-service developer to look at our logging library and immediately see the benefits and want to start using it. The next section will show how TDD benefits the most from an awareness of who will be using the logging library.

How will TDD help build a logging library?

The biggest benefit we'll get from using TDD to build a logging library is the customer focus. It's easy to make assumptions about how something should be designed, especially when there are already many common solutions that do something similar.

It's like a trail that's easy to follow. The trail might take you someplace you want to go or maybe nearby. And if the destination is vague or unknown, then a trail becomes even easier to follow. But if you know exactly where you want to go, then you can follow a trail when convenient and leave the trail when it no longer takes you where you want to go.

TDD encourages us to think about how we want to use the software that we're building. This, in turn, lets us customize the solution to best fit our needs. In other words, it lets us know when to leave the trail and start a new path.

We also benefit from having tests that can verify the behavior of the software because building software is not like walking along a path one time. We don't start at the beginning and walk directly to the destination or final software that we want to build.

Instead, we refine the path. It's more like starting with a map and drawing a rough path to the destination. Maybe the path aligns with known and existing paths, and maybe not. Once we have the rough path drawn, we refine it by adding more details. Sometimes the details might make us change our path. By the time we're done, we've walked the path so many times we've lost count.

TDD helps by guiding us to a simple solution first and verifying that it works. This is like the rough path initially drawn on the map. Each time we enhance and refine the solution, we have tests to make sure that we don't break anything. This is like checking to make sure we are still on the path.

Sometimes, an enhancement results in the need for a design change. This is like discovering a river blocking our way that didn't appear on the initial map. We need to go around it by finding a different place to cross. The path changes, but the need to stay on the new path remains. This is where the tests help us. We have a new design that fixes an unforeseen problem, but we can still use the existing tests to validate the new design. We have confidence in the new design because it fixes the unexpected problem while still performing as expected.

TDD will help increase the usage of the logging library by other projects by providing clear and documented examples of how to use the library. This would be like producing videos of walking along the path so that future travelers will know what to expect and whether they want to follow the path before starting. Given a choice between our fully documented and easy-to-follow examples versus a similar library that claims to offer the same results but without proof, most people will prefer our library. Over time, our solution will get even better as we incorporate feedback from users.

Let's start by thinking about the destination we want. This would be the ideal logging library for our intended customers. What would that look like? This will form the first rough path that we'll start to refine.

What would the ideal logging library look like?

When designing software, it's good to remember that you don't have to design something completely new if common solutions already meet your needs exactly. A new design that's different just for the sake of being different is just going to confuse people. A new design that's different because the existing designs don't quite work and the differences solve a real need is good. So before we begin dreaming up new logging designs, let's first look at common ideas and see if we really need something new.

To be completely thorough, we should also try to use what C++ already provides. Maybe that is enough and we don't need a library. Let's say we have the following code that tries to calculate the result of starting with 1 and doubling the value three times. The correct answer should be 8, but this code has a bug:

```cpp
#include <iostream>

int main ()
{
    int result = 1;
    for (int i = 0; i <= 3; ++i)
    {
        result *= 2;
    }
    std::cout << "result=" << result << std::endl;

    return 0;
}
```

It prints a result of 16 instead of 8. You probably already see the problem, but let's imagine that the code is a lot more complicated, and the problem was not obvious. As the code is currently written, we'll go through the loop four times instead of just three times.

We could add extra output such as this to help find the problem:

```cpp
    for (int i = 0; i <= 3; ++i)
    {
        std::cout << "entering loop" << std::endl;
        result *= 2;
    }
```

Which, when run, produces the following output:

```
entering loop
```

```
entering loop
entering loop
entering loop
result=16
```

The results clearly show that the loop was run four times instead of just three.

What we just did by adding extra text to the output that showed what was happening while running the code is the core of logging. Sure, we can add more features or do more. But the questions we need to ask are is this enough, and will this meet our needs?

Using `std::cout` is not enough and will not meet our needs for several reasons:

This simple example already used the console output to display the results. Normally, services should avoid sending text to the console because there won't be anybody watching the screen for the results.

Even if the console was the desired destination for the program output, we shouldn't mix the extra logging output with the regular output.

We could send the logging output to a different stream, such as `std::cerr`, but this is not a full solution. Is the logging output always an error? Maybe it helps us identify that our program is actually running correctly and that the problem must be somewhere else.

Logging extra information is useful, but not all the time. Sending output directly to `std::cout` doesn't give us a way to turn it off without changing the source code and rebuilding.

And it would be nice if the logging output included extra information, such as the date and time. We could add this extra information, but then we'd have to add it every time we called `std::cout` to log information.

We're making progress in the design because we've just eliminated one possible path. It's always good to think about what you already have available before looking for a solution elsewhere.

What if we put the logging into a function and called that function instead of using `std::cout` directly? The code might look like this:

```
#include <fstream>
#include <iostream>

void log (std::string_view message)
{
    std::fstream logFile("application.log", std::ios::app);
    logFile << message << std::endl;
}
```

```
int main ()
{
    int result = 1;
    for (int i = 0; i <= 3; ++i)
    {
        log("entering loop");
        result *= 2;
    }
    std::cout << "result=" << result << std::endl;

    return 0;
}
```

This is a big improvement already. Even though the application is still using `std::cout` to display the result, we're not adding to the noise with even more console output for logging. Now the logging output goes into a file. This also avoids mixing the logging with the regular results.

We could even add a check inside the `log` function to see if the message should be logged or ignored. And having everything wrapped up in a function would also make it easy to add common information such as the date and time.

Is the ideal solution just a function?

Not really, because we also need to configure the logging. The code just shown was very simple and used a fixed log filename. And other features are missing that will improve the logging experience, for example:

- The code currently opens and closes the log file for each message.

- There is an assumption that the message should go to a file. Maybe we want the message to go somewhere else or to a file, in addition to somewhere else.

- The message is a single text string.

- The code does not handle multiple threads trying to log messages at the same time.

- The log function makes the main application wait until the log message has been written before the application can proceed.

- We haven't added anything specific to our desired customer, the micro-service developer, such as the ability to filter messages.

What we have is a good start, confirming that there are enough requirements to justify the need for a library.

Looking at other similar solutions is also a good idea, and there are many. One well-known logging library comes from the Boost C++ libraries and is called Boost.Log. This library allows you to start logging to the console in a simple way. And the library is extensible and fast. It's also big. Even though it starts out simple, I've spent days reading through the documentation. One thing leads to another and before I knew it, I was learning about other technologies that the logging library uses.

While the Boost.Log library might start out simple, it can quickly require you to learn a lot more than expected. I'd like to create something that stays simple to use. Our ideal logging library should start out simple to use and hide any necessary complexity so that the user is not buried in options. We're not trying to build a logging library that can do everything. We have a specific user in mind and will use TDD to focus on the needs of that micro-service developer.

The next section will begin the process of creating a logging library. Before we begin writing tests, we'll need to create a new project, which the next section explains.

Starting a project using TDD

Now that we've determined that a logging library is a good idea and is justified, it's time to start a new project. Let's start with the `log` function from the previous section and create a new project. The `log` function looks like this:

```
void log (std::string_view message)
{
    std::fstream logFile("application.log", std::ios::app);
    logFile << message << std::endl;
}
```

Where will we put this `log` function, and what will the test project structure look like? In the earlier chapters, we tested the unit test library. This is the first time we'll be using the unit test library as something outside the actual project we're working on. The project structure will look like this:

```
MereMemo project root folder
    MereTDD folder
        Test.h
    MereMemo folder
        Log.h
        tests folder
            main.cpp
            Construction.cpp
```

The new structure uses a containing folder called the MereMemo project root folder. Like the unit test library is called MereTDD, I decided to continue the theme with the word mere and call the logging library MereMemo. Other choices were already in use, and the word memo represents the idea of writing something down to remember it.

You can see that inside the root folder is a folder called MereTDD with only the Test.h file. We no longer need to include the tests for the unit test library. We're going to use the unit test library now instead of developing it further. If we ever do need to make changes to the unit test library, then we'll go back to the previous project that contains the tests for the unit test library.

The project root folder gives us a place to put both the unit test library header file in its own folder and the logging library also in its own folder.

Inside the MereMemo folder is a file called Log.h. This is where we'll put the log function. And there is also a folder called tests, which will contain the unit tests for the logging library. Inside the tests folder is where we will find the main.cpp file and all the other unit test files. For now, there is just one unit test file called Construction.cpp, which is empty and doesn't contain any tests yet.

I should also mention that you don't need to put the MereTDD folder inside your project root folder like this. You can put it anywhere you want. This is like installing the unit test library on your computer. Since the unit test library is really just a single header file, there isn't anything that needs to be installed. The header file just needs to exist somewhere on your computer in a convenient place so that you know the path. We'll need to add the path to the project settings in your development tools so that the compiler knows where to find Test.h. I'll explain more about this step in just a moment.

We need the usual include guards in Log.h, and after putting the log function inside, then Log.h should look like this:

```
#ifndef MEREMEMO_LOG_H
#define MEREMEMO_LOG_H

#include <fstream>
#include <iostream>
#include <string_view>

namespace MereMemo
{

inline void log (std::string_view message)
{
    std::fstream logFile("application.log", std::ios::app);
    logFile << message << std::endl;
```

```
}

} // namespace MereMemo
```

```
#endif // MEREMEMO_LOG_H
```

The `log` function now needs to be inline since it is in its own header file and could be included multiple times in a project.

We can mostly copy the contents of `main.cpp` from the unit test library project and use it to run the unit tests for the logging library project too. We need to make a minor change to how we include `Test.h`, though. The `main.cpp` file should look like the following example:

```
#include <MereTDD/Test.h>

#include <iostream>

int main ()
{
    return MereTDD::runTests(std::cout);
}
```

You can see that we now include `Test.h` using angle brackets instead of quotation marks. This is because `Test.h` is not directly part of the logging library; it's now a file being included from another project. The best way to include files from other projects or libraries is to keep them separate in their own folders and change your project settings in your development tools to tell the compiler where to find the files needed.

For my development work, I'm using the CodeLite **integrated development environment (IDE)**, and the project settings are available by right-clicking on the project and choosing the **Settings** menu option. Inside the pop-up dialog, there is a section for the compiler settings. And on the compiler settings page, there's an option to specify the include paths. CodeLight has some predefined paths that can be used to identify things such as the path to the current project. I set the include paths to look like this:

```
.;$(ProjectPath)
```

The include paths are separated by semicolons. You can see there are two paths specified. The first is a single dot, which means to look for included files in the current folder. This is how the project-specific include files that use quotation marks are found. But I also added a path with the special syntax using the dollar sign and parentheses, which tells CodeLite to look in the project root folder for additional include files. What actually happens is that CodeLite interprets the paths, including its special predefined paths, such as ProjectPath, and sends the real filesystem paths to the compiler. The compiler doesn't know anything about `ProjectPath`, parentheses, or dollar signs.

If you decide to place the unit test library somewhere else on your computer, you will need to add the full path instead of using ProjectPath. And if you're using a different IDE instead of CodeLite, then the process will be similar. All IDEs have their own way of specifying the include paths to be used by the compiler. The settings are almost always in a settings dialog that you can open from the project.

With all this setup and project configuration work done, it's time to start writing some tests, beginning in the next section.

Logging and confirming the first message

Now that we have a project ready, we can begin writing some tests and designing the logging library. We already created an empty unit tests file called Construction.cpp. I like to start with some simple tests that ensure classes can be constructed. We can also use this to make sure that simple functions can be called. This section will focus on creating a single test to log our first message and confirm that it all works.

We already have the log function from earlier, which opens a file and appends a message. Let's add a test that calls log and writes something. The following example shows how to edit Construction. cpp to add the first test:

```
#include "../Log.h"

#include <MereTDD/Test.h>

TEST("Simple message can be logged")
{
    MereMemo::log("simple");
}
```

Because we're testing the logging library, we need to include Log.h, which is found in the parent directory where Construction.cpp is located. We use quotation marks for Log.h because it's in the same project. Later, if you want to use the logging library for your own project, you'll just need to put Log.h in a known location and include it with angle brackets, just like how we now include Test.h with angle brackets.

The single test just calls the log function. This really just reorganizes the code we started in this chapter by creating a real project and using tests instead of directly writing code in main. Building and running the project shows the following output to the console:

```
Running 1 test suites
-------------- Suite: Single Tests
------- Test: Simple message can be logged
```

```
Passed
-----------------------------------
Tests passed: 1
Tests failed: 0
```

The single test ran and passed. But it will pass no matter what as long as it doesn't throw an exception. That's because we don't have any confirmations in the test. The real output that we're interested in doesn't even appear in the console. Instead, it all goes to the log file called `application.log`. When I run the project from the CodeLite IDE, it shows a similar output. But it seems that CodeLite runs the code from a temporary folder. Other IDEs also do something similar, and it can be hard sometimes to keep up with the temporary locations. So instead, you might want to use your IDE to build the project and then open a separate terminal window to run the test application manually. This way, you are in full control over where the application is run and have a window open that you can use to examine the log file that gets created.

My application gets built in a folder called `Debug`, and the contents of that folder after building contain the executable file and the object files that are used to create the final executable. There is no file called `application.log` until the test application project is run. Once the project is run, the `application.log` file can be printed to the console like this:

```
$ cat application.log
simple
```

At the `$` prompt, the `cat` command is used to display the contents of the `application.log` file, which contains a single line with a simple message. If we run the project again, then we'll append new content to the same log file, which looks like this:

```
$ cat application.log
simple
simple
```

After running the application twice, we get two messages in the log file. Both messages are identical, which is going to make it hard to determine whether anything new was added to the log file or not. We need a way to create unique messages and then a way to verify that a particular message is found in the log file. This will let us add a confirmation to the test to verify that a message was logged without needing to manually examine the log file each time the test is run.

Other tests might need the ability to generate unique messages and then verify the contents of the log file, and we might want to put these other tests in different `test.cpp` files. So that means we should add a helper file to write the required code so that it can be shared with other tests in other files.

A common name for a helper file like this is `Util`. It seems like every project has `Util.h` and `Util.cpp`, and this is why. It's a good place to put useful code that can be shared throughout a project.

What would the test look like if we had these helper functions? Change `Construction.cpp` to look like the following screenshot:

```cpp
#include "../Log.h"

#include "Util.h"

#include <MereTDD/Test.h>

TEST("Simple message can be logged")
{
    std::string message = "simple ";
    message += Util::randomString();

    MereMemo::log(message);

    bool result = Util::isTextInFile(message,
        "application.log");
    CONFIRM_TRUE(result);
}
```

We need to include `Util.h`, and then we can make the message unique by appending a random string we get from calling `randomString`. The full message is stored in a variable so that we can use it when logging and verifying. After logging the message, we call the other new function, `isTextInFile`, to verify.

One problem with this test is the need to specify the log filename. Right now, the log filename is hardcoded in the `log` function. We're not going to fix the log filename problem right away. With TDD, we take things one step at a time. If you have an issue tracking system, adding the log filename problem to the tracking system will make sure it isn't forgotten.

Now that we have an idea of how the utility functions will be used, let's add both utility files to the project in the `tests` folder and add a couple of function declarations in `Util.h` like this:

```cpp
#ifndef MEREMEMO_TESTS_UTIL_H
#define MEREMEMO_TESTS_UTIL_H

#include <string>
#include <string_view>
```

```
struct Util
{
    static std::string randomString ();

    static bool isTextInFile (
        std::string_view text,
        std::string_view fileName);
};


#endif // MEREMEMO_TESTS_UTIL_H
```

The first function will let us generate a random string that we can use to make the messages unique. There's always a possibility that we'll get duplicate strings, which could cause us to think that a log file contains a new log message when, in fact, we see a previous message that used the same random string. In practice, this shouldn't be a problem because we won't just be logging random strings. We'll be adding random strings to other text that will be unique for each test.

When I was originally developing this code, I used the same text for many of the tests. The random number added to the end makes each message unique. Or, at least, I didn't notice any duplicate messages. The tests all worked great until I got to *Chapter 15, How to Test With Multiple Threads*, and added 150 new messages in a single test. The problem wasn't due to multiple threads. The problem was always a possibility and didn't appear until the extra messages increased the probability of duplicate messages. We're going to avoid the problem by using unique base message text for each test.

The second function will let us confirm that some text actually exists in a file, and we can use this to verify that a particular message exists in the log file.

I like to make functions like this static methods in a struct. This helps to make sure that the implementations match the declarations in the header file. The implementation goes in Util.cpp like this:

```
#include "Util.h"

#include <fstream>
#include <random>

std::string Util::randomString ()
{
    return "1";
}


bool Util::isTextInFile (
```

```
        std::string_view text,
        std::string_view fileName)
    {

        return false;
    }
```

The implementations don't do anything other than return for now. But this lets us build and verify that the test fails.

Why would we want to make sure that the test fails?

Because it helps to validate a passing result once we actually implement the functions. A failure ensures that the test is written correctly and can catch a failure. Once we implement the helper functions, then the test will pass, and we can be sure that the pass is coming from the helper implementation and not just a test that will always pass anyway.

The following screenshot shows the expected failure when running the project:

```
Running 1 test suites
-------------- Suite: Single Tests
------- Test: Simple message can be logged
Failed confirm on line 15
    Expected: true
---------------------------------
Tests passed: 0
Tests failed: 1
```

Even though a manual check of the application.log file shows that the expected message did get written to the end of the log file:

```
$ cat application.log
simple
simple
simple 1
```

Let's fix the randomString function now to ensure we can get unique messages logged. We'll need to include chrono to be able to set up a random number generator with a seed based on the current time. The following code snippet shows the relevant code in Util.cpp:

```
#include <chrono>
#include <fstream>
#include <random>
```

```cpp
std::string Util::randomString ()
{
    static bool firstCall = true;
    static std::mt19937 rng;
    if (firstCall)
    {
        // We only need to set the seed once.
        firstCall = false;

        unsigned int seed = static_cast<int>(
            std::chrono::system_clock::now().
            time_since_epoch().count());
        rng.seed(seed);
    }
    std::uniform_int_distribution<std::mt19937::result_type>
dist(1, 10000);

    return std::to_string(dist(rng));
}
```

Because this uses random numbers, you'll get different results each time. The test still fails, but after running the test application a couple more times, my `application.log` file looks like this:

```
$ cat application.log
simple
simple
simple 1
simple 2030
simple 8731
```

The message is now somewhat unique, with a slight chance of duplicate log messages. That's good enough for now, and we can move on to getting the verification function working. We've been keeping the log file between test application runs, so the new message is appended each time. A real test run to verify your code should start with a clean environment with no leftover files from a previous run.

I'm showing the code for both the random string and the verification without fully explaining everything. That's because random numbers and file searches are needed but are not completely in scope for explaining TDD. It's easy to get off topic by explaining all the details about random numbers or even searching text files for matching strings.

The implementation of the `isTextInFile` function looks like this:

```cpp
bool Util::isTextInFile (
    std::string_view text,
    std::string_view fileName)
{
    std::ifstream logfile(fileName.data());
    std::string line;
    while (getline(logfile, line))
    {
        if (line.find(text) != std::string::npos)
        {
            return true;
        }
    }
    return false;
}
```

All this function does is open the log file, read each line, and try to find the text. If the text is found, then it returns true, and the function returns false if it cannot find the text in any line.

Building and running the project now shows the test passes like this:

```
Running 1 test suites
-------------- Suite: Single Tests
------- Test: Simple message can be logged
Passed
----------------------------------
Tests passed: 1
Tests failed: 0
```

We now have a way to write log messages to a file and confirm that the message appears in the file. The code could be more efficient because, right now, it searches through the entire log file starting at the beginning when looking for the text. But our goal is not to write the best log file searching tool. The test log files will not likely grow too big, so a simple approach to searching should work well.

A log needs more than just messages to be useful. The following section will add timestamps to the messages, adding the minimum set of features needed for a logging library.

Adding timestamps

A single log file might provide some value with just messages, but recording the date and time of each log message makes it much more valuable. And once we start working with multiple log files from different micro-services, the need to order the log messages by time becomes critical.

It's fairly simple to add timestamps. All we need to do is get the current time, format it into a standard timestamp that eliminates misunderstanding between the year, month, and day, and then send the timestamp to the log along with the message. The caller doesn't need to do anything.

It's also difficult to test directly to make sure it works. We can manually open the log file and see the timestamps; that will be enough for now. We're not going to add any new tests for the timestamps.

All we need to do is modify Log.h so it looks like this:

```cpp
#include <chrono>
#include <ctime>
#include <fstream>
#include <iomanip>
#include <iostream>
#include <string>
#include <string_view>

namespace MereMemo
{

inline void log (std::string_view message)
{
    auto const now = std::chrono::system_clock::now();
    std::time_t const tmNow =
        std::chrono::system_clock::to_time_t(now);
    auto const ms = duration_cast<std::chrono::milliseconds>(
        now.time_since_epoch()) % 1000;

    std::fstream logFile("application.log", std::ios::app);
    logFile << std::put_time(std::gmtime(&tmNow),
            "%Y-%m-%dT%H:%M:%S.")
            << std::setw(3) << std::setfill('0')
            << std::to_string(ms.count())
            << " " << message << std::endl;
}
```

We need to include some new system headers, `chrono`, `ctime`, `iomanip`, and `string`. A better way to format dates and times involves a new system header file called `format`. Unfortunately, even though `format` is part of C++20, it's still not widely implemented by most standard libraries. So this code uses a slightly older way to do the formatting that uses the standard `put_time` function.

We start by getting the system time. Then we need to convert the time to an older time format called `time_t`. Even though I often only mention time, I usually mean both time and date.

We want the timestamps to be fairly precise and just logging the time down to the seconds is not nearly precise enough. So, we need to cast the time to milliseconds and divide by 1,000 to get the fraction of a second we need.

The function continues to open the log file as before. But now, it calls `put_time` with the **Coordinated Universal Time** (**UTC**) or **Greenwich Mean Time** (**GMT**). Both terms mean almost the same thing for most practical purposes. GMT is an actual time zone, while UTC is a standard. They are normally the same and are used instead of local time to avoid issues with different time zones. By calling `gmtime`, we can ensure that logs generated on machines in different time zones will be ordered correctly because all the times refer to the same time zone.

If you're using the Visual Studio tools from Microsoft, you'll likely get an error with the use of `gmtime`. I mentioned this is an older solution, and some compilers will complain that `gmtime` could be unsafe. The recommended replacement is `gmtime_s`, but this replacement function requires some extra checks in the code to see if it is available. Other compilers might also complain about `gmtime` and normally tell you in the error message what needs to be done to fix the problem. The error message from Visual Studio says that if we want to use `gmtime`, we need to define `_CRT_SECURE_NO_WARNINGS` in the project settings under C++ preprocessor definitions.

The strange formatting involving percent signs and uppercase and lowercase letters tells `put_time` how to format the various elements of the date and time. We want to format the date and time according to the **ISO-8601** standard. The most important part is that the standard says that the year comes first with four digits followed by the two-digit month, and then the two-digit day. Dashes are allowed between the numbers.

Without a standard like this, a date such as 10-07-12 could mean different dates to different people. Is that October 7, 2012? Or July 10, 2012? Or July 12, 2010? Or December 7, 2010? About the only thing we can all agree on is that the year is probably not 2007. Even using a four-digit year still leaves room for the month and day to get mixed up. By using ISO-8601, we all agree that the year is first, then the month, and then the day.

Next in the standard is the capital letter T. All this does is separate the date from the time portion. The time comes next and is much less confusing because we all agree that the hour comes first, then the minutes, and then the seconds. We put a dot after the seconds before displaying the fractional milliseconds.

After making these changes, deleting the old log file, and building and running the project a few times, we can see that the log file looks something like this:

```
$ cat application.log
2022-06-13T03:37:15.056 simple 8520
2022-06-13T03:37:17.288 simple 1187
2022-06-13T03:37:18.479 simple 2801
```

The only thing we're not doing is including specific text in the timestamp that shows the time is UTC. We could add specific time zone information, but this should not be needed.

We have timestamps and can log a single string of text. The next section will let us log more than a single piece of text.

Constructing log messages with streams

Having a `log` function that accepts a single string to display is not the easiest function to use. Sometimes, we might want to log more information. We might also want to log different types and not just strings. This is where we can use the powerful streaming ability used extensively in C++.

We're already using a stream inside the `log` function. All we need to do to give the full power of the stream to the caller is to stop sending the single message text to the stream inside the `log` function and instead return the stream itself to the caller. The caller will then be free to stream whatever is needed.

We can see what this will look like by first modifying the test; we have to use the `log` function as if it returned a stream. The modified test looks like this:

```
TEST("Simple message can be logged")
{
    std::string message = "simple ";
    message += Util::randomString();
    MereMemo::log() << message << " with more text.";

    bool result = Util::isTextInFile(message,
        "application.log");
    CONFIRM_TRUE(result);
}
```

We now call the `log` function without any arguments. Instead of passing the `message` variable as an argument, we use the `log` function's return value as a stream and directly send the message along with another piece of text to the stream.

Notice how we need to make sure to add a space before the second piece of text. This is to prevent the text from joining up with the message that comes before it. As with any stream, it's up to the caller to ensure the text doesn't run together.

There is a slight problem we need to work out. Previously, when we were handling the message inside the `log` function, we were able to add a newline to the end of the message. We still want log messages to appear on their own lines in the log file. But we also don't want the caller to always have to remember to add a newline.

One of our goals is to make this logging library easy to use. Requiring each call to `log` to include a newline at the end makes the use tedious. So instead, a simple temporary solution is for the `log` function to add a newline to the *beginning* of each log message.

This has a strange side effect. The very first line in a log file will be empty. And the last line will not have a newline. But the overall effect is still that each log message will appear on its own line. And that's the behavior that we want to keep. This temporary solution will be fixed properly in the following chapter.

The change to the `log` function in `Log.h` is also simple and looks like this:

```
inline std::fstream log ()
{
    auto const now = std::chrono::system_clock::now();
    std::time_t const tmNow =
        std::chrono::system_clock::to_time_t(now);
    auto const ms = duration_cast<std::chrono::milliseconds>(
        now.time_since_epoch()) % 1000;

    std::fstream logFile("application.log", std::ios::app);
    logFile << std::endl
        << std::put_time(std::gmtime(&tmNow),
        "%Y-%m-%dT%H:%M:%S.")
        << std::setw(3) << std::setfill('0')
        << std::to_string(ms.count())
        << " ";

    return logFile;
}
```

Instead of returning `void`, this new version returns `std::fstream`. You can see that the first thing that gets sent to the log file stream is `std::endl`, which ensures each log message gets its own line in the log file. The entire idea of returning `std::fstream` is a temporary solution that will be enhanced in the next chapter.

Then, instead of sending a message to the stream, the function returns the stream after sending the timestamp. This lets the caller send whatever values are needed to the stream. An empty space is added after the timestamp to make sure that the timestamp doesn't run into additional text that gets streamed by the caller.

An interesting thing to consider about the return type is what will happen to fstream once the function returns. We construct fstream inside the `log` function and then return the stream by value. Returning a stream by value was not possible before C++11, and we can do this only because we now have the ability to *move* the stream out of the `log` function. The code doesn't need to do anything special to enable the move. It just works with modern C++. We will run into the move issue again in the next chapter when we fix the temporary solution with newlines.

Following TDD to design software encourages working solutions that get enhanced instead of trying to come up with a perfect design in the beginning. I can tell you from experience that it's not possible to think of every little design issue ahead of time. Adjustments need to be made along the way, which tends to turn a perfect design into something less than it was. I like TDD because designs start with the end user in mind and the enhancements work around small issues such as our newline problem. The end result is better than what it was in the beginning. Following TDD lets the design stay true to the things that matter the most while being flexible where needed.

We still need to consider what happens to the stream once it leaves the `log` function. The test code is not storing the stream in a local variable, so it will get destroyed. But it will only get destroyed at the end of the expression in which it was created. Let's say we call `log`, as follows:

```
MereMemo::log() << message << " with more text.";
```

The `std::fstream` that gets returned from the `log` function remains valid until the semicolon at the end of the line. The lifetime of the stream needs to remain valid so that we can continue to use it to send the message and the additional text.

Building and running the project a few times shows that the single test passes each time. And the log file contains the extra text, as displayed in the following screenshot:

```
$ cat application.log

2022-06-13T05:01:56.308 simple 5586 with more text.
2022-06-13T05:02:02.281 simple 2381 with more text.
2022-06-13T05:02:05.621 simple 8099 with more text.
```

You can see the empty line at the beginning of the file. But each message is still on its own line, and the caller can now log other information. Let's create a new test to make sure. The new test will look like the following example:

```
TEST("Complicated message can be logged")
{
    std::string message = "complicated ";
    message += Util::randomString();
    MereMemo::log() << message
```

```
            << " double=" << 3.14
            << " quoted=" << std::quoted("in quotes");

    bool result = Util::isTextInFile(message,
    "application.log");
    CONFIRM_TRUE(result);
}
```

This new test logs a double literal value directly and even logs the result of calling the std::quoted function. The quoted function just puts quotation marks around what is given to it. Even though the text "in quotes" already looks like it has quotation marks, remember that these marks are for the compiler to know when the text begins and ends. The quotation marks in the source code are not actually part of the text, just like how the quotation marks from the other string literals such as "double=" don't appear in the log message. But because we call std::quoted, we will get quotation marks in the output.

The interesting thing about std::quoted is that the return value can really only be used to send to a stream. The actual type is undefined by the C++ standard, and the only requirement is that it can be sent to a stream.

Building and running the project shows that both tests pass. The following example shows what the application.log file looks like after deleting it and running the tests a couple of times:

```
$ cat application.log

2022-06-13T05:47:36.973 simple 6706 with more text.
2022-06-13T05:47:36.975 complicated 1025 double=3.14 quoted="in
quotes"
2022-06-13T05:47:39.489 simple 4411 with more text.
2022-06-13T05:47:39.495 complicated 9375 double=3.14 quoted="in
quotes"
```

We now have the ability to create log messages with timestamps, save them to a log file, and can send whatever data we want to each log message. The usage is simple and intuitive for C++ developers who are already familiar with sending information to a stream such as std::cout.

Summary

In this chapter, you learned how to use the unit test library to begin a new project using TDD. Even though we only have two tests, we already have a working logging library that is easy to use and understandable by any C++ developer.

The two tests will help ensure we don't break the simple design started in this chapter as we extend the logging library in later chapters. The next chapter, in particular, will extend the logging library to better fit the needs of our intended user, a micro-services developer. We'll be adding the ability to tag log messages and then use the tags to enable powerful filtering options.

The TDD Process in Depth

We're going to add a lot of code to the logging library in this chapter, and while that's good, it's not the main purpose of the chapter.

This is a chapter about the **test-driven development** (TDD) process. Wasn't *Chapter 3* also about the TDD process? Yes, but think of the earlier chapter as an introduction. This chapter will explore the TDD process in detail with a lot more code.

You'll get ideas for writing your own tests, how to figure out what's important, and how to refactor code without rewriting tests too, and you'll also learn when testing is too much and learn about many different types of tests.

The basic TDD process remains as follows:

- To write tests first that use the software in a natural and intuitive way
- To get the code building with minimal changes even if we need to provide fake or stubbed-out implementations
- To get basic scenarios working
- To write more tests and enhance the design

Along the way, we'll add log levels, tags, and filtering to the logging library.

Specifically, we'll cover the following main topics in this chapter:

- Finding gaps in the testing
- Adding log levels
- Adding default tag values
- Exploring filtering options
- Adding new tag types
- Refactoring the tag design with TDD

- Designing tests to filter log messages

- Controlling what gets logged

- Enhancing filtering for relative matches

- When is testing too much?

- How intrusive should tests be?

- Where do integration or system tests go in TDD?

- What about other types of tests?

Technical requirements

All code in this chapter uses standard C++ that builds on any modern C++ 20 or later compiler and standard library. The code uses the testing library from *Part 1* of this book, *Testing MVP*, and continues the development of a logging library started in the previous chapter.

You can find all the code for this chapter in the following GitHub repository:

`https://github.com/PacktPublishing/Test-Driven-Development-with-CPP`

Finding gaps in the testing

We really need more tests. Right now, we only have two logging tests: one for simple log messages and the other for more complicated log messages. The two tests look like this:

```
TEST("Simple message can be logged")
{
    std::string message = "simple ";
    message += Util::randomString();
    MereMemo::log() << message << " with more text.";

    bool result = Util::isTextInFile(message,
    "application.log");
    CONFIRM_TRUE(result);
}

TEST("Complicated message can be logged")
{
    std::string message = "complicated ";
    message += Util::randomString();
```

```
    MereMemo::log() << message
        << " double=" << 3.14
        << " quoted=" << std::quoted("in quotes");

    bool result = Util::isTextInFile(message,
    "application.log");
    CONFIRM_TRUE(result);
}
```

But is there a good way to find more tests? Let's look at what we have so far. I like to start with simple tests. Can things be constructed?

That's why the two tests we have so far are in a file called Contruction.cpp. This is a good place to start when looking for gaps in your testing. Do you have a simple test for each thing that you can construct? Normally, these will be classes. Write a test for each constructor of each class that your project provides.

For the logging library, we don't have any classes yet. So instead, I created a simple test that calls the log function. Then, another test calls the same function in a slightly more complicated way.

There is an argument to be made that the complicated test duplicates some of the functionality of the simple test. I think what we have so far is okay, but it is something you should be aware of to avoid having a test that does everything another test does plus a little more. As long as a simple test represents a common use case, then it's valuable to include it, even if another test might do something similar. In general, you want tests that will capture how your code will be used.

Other things to think about when looking for gaps in your testing can be found by looking for symmetry. If you have construction tests, maybe you should consider destruction tests. We don't have anything like that for the logging library—at least, not yet—but it is something to consider. Another example of symmetry can be found later in this chapter. We'll need to confirm that some text exists in a file. Why not include a similar test that makes sure some different text does *not* exist in the file?

Major features are a good source for tests. Think about which problems your code solves and write tests that exercise each feature or capability. For each feature, create a simple or common test, and then consider adding a more complicated test, some error tests that explore what can go wrong, and some tests that explore more purposeful misuses to make sure your code handles everything as expected. You'll even see one example in the next section where a test is added just to make sure that it compiles.

This chapter will mostly explore tests for missing features. We're just getting started with the logging library, so the majority of new tests will be based on new features. This is common for a new project and is a great way to let the tests drive the development.

The next section will add a new feature by first creating tests to define the feature.

Adding log levels

Logging libraries have a common idea of *log levels* that let you control how much information gets logged when an application is run. Let's say you identify an error condition that needs a log message. This error should almost always be logged, but maybe there's another place in the code where you decide that it might be useful to record what is happening. This other place is not always interesting, so it would be nice to avoid seeing those log messages all the time.

By having different log levels, you can decide how verbose your log files become. There are a couple of big problems with this approach. The first thing is simply defining what the log levels should be and what each should mean. Common log levels include errors, warnings, general informational messages, and debugging messages.

Errors tend to be easy to identify unless you also want to split them into normal errors and critical errors. What makes an error critical? Do you even need to tell the difference? In order to support as many different customers as possible, a lot of logging libraries provide different log levels and leave it up to the programmer to figure out what each level means.

The logging levels end up being used primarily to control how much information gets logged, which can help reduce the size of the log files when the application is running without any problems or complaints. This is a good thing but it leads to the next big problem. When something needs further investigation, the only way to get more information is to change the logging level, rerun the application, and hope to catch the issue again.

For large applications, changing the logging level to record more information can quickly cause so much extra information that it becomes difficult to find what you need. The additional log messages can also fill up storage drives and cause extra financial charges if the log files are sent to vendors for further processing. The debugging process is usually rushed so that the new logging level is in effect for a short period of time.

To get around the need to change the logging level for the entire application, a common practice is to temporarily change the level that specific parts of the code use when logging information. This requires that an application be rebuilt, deployed, and then put back once the problem is found.

How does all this discussion about logging levels help us design a logging library? We know who our target customer is: a microservice developer who will likely be working with large applications that can produce large log files. Thinking about what would help your customer the most is a great way to create a design.

We're going to fix the two big problems identified. First, we're not going to define any logging levels in the logging library. There will not be any notion of an error log message versus a debug log message. This doesn't mean that there will be no way to control how much information will get logged, just that the whole idea of using log levels is fundamentally broken. The levels themselves are too confusing, and turning them on and off leads quickly to information overload and rushed debugging sessions.

The idea of adding extra information such as a log level to a log message is good. If we come up with a general-purpose solution that can work for log levels as well as other attached information, then we can let the user add whatever is needed and makes sense. We can provide the ability to add log levels without actually defining what those levels will be and what they mean.

So, the first part of the solution will be a general-purpose *tagging* system. This should avoid the confusion of fixed log levels that are defined by the library. We'll still refer to the idea of a log level, but that's only because the idea is so common. However, our log levels will be more like log-level tags because there won't be any idea of one log level being above or below another log level.

The second part will need something new. The ability to control whether a message is logged or not based on the value of a log-level tag will just lead to the same problem as before. Turning a log level on will end up opening the logs everywhere and still lead to a flood of extra log messages. What we need is the ability to finely control what gets logged instead of turning extra logging on or off everywhere. We'll need the ability to filter on more than just a log level.

Let's take these two ideas one at a time. What would a general tagging system look like? Let's write a test to find out! We should create a new file called Tags.cpp in the tests folder that looks like this:

```
#include "../Log.h"

#include "LogTags.h"
#include "Util.h"

#include <MereTDD/Test.h>

TEST("Message can be tagged in log")
{
    std::string message = "simple tag ";
    message += Util::randomString();
    MereMemo::log(error) << message;

    std::string taggedMessage = " log_level=\"error\" ";
    taggedMessage += message;
    bool result = Util::isTextInFile(taggedMessage,
        "application.log");
    CONFIRM_TRUE(result);
}
```

The most important part of this test is the `log` function call. We want this to be easy to use and to quickly convey to anybody reading the code that a tag is involved. We don't want the tag to get hidden in the message. It should stand out as different and at the same time not be awkward to use.

The confirmation is a little more complicated. We want the output in the log file to use a `key="value"` format. This means that there is some text followed by an equals sign and then more text inside quotation marks. This format will let us easily find tags by looking for something like this:

```
key="value"
```

For the log level, we'll expect the output to look like this:

```
log_level="error"
```

We also want to avoid mistakes such as spelling or differences in capitalization. That's why the syntax doesn't use a string, which could be mistyped like this:

```
MereMemo::log("Eror") << message;
```

By avoiding strings, we can get the compiler to help make sure that tags are consistent. Any mistake should result in a compilation error instead of a malformed tag in the log file.

And because the solution uses a function argument, we don't need to provide special forms of `log` such as `logError`, `logInfo`, or `logDebug`. One of our goals was to avoid defining specific logging levels in the library itself and instead come up with something that will let the user decide what the log levels will be, just like any other tag.

This is also the reason for the extra include of `LogTags.h`, which is also a new file. This is where we will define which log levels we will use. We want the definition to be as simple as possible because the log library will not define these. The `LogTags.h` file should be placed in the `tests` folder and look like this:

```
#ifndef MEREMEMO_TESTS_LOGTAGS_H
#define MEREMEMO_TESTS_LOGTAGS_H

#include "../Log.h"

inline MereMemo::LogLevel error("error");
inline MereMemo::LogLevel info("info");
inline MereMemo::LogLevel debug("debug");

#endif // MEREMEMO_TESTS_LOGTAGS_H
```

Just because the logging library doesn't define its own logging levels doesn't mean it can't help with this common task. We can make use of a helper class that the library will define called `LogLevel`. We include `Log.h` in order to gain access to the `LogLevel` class so that we can define instances. Each instance should have a name such as `error` that we will use when logging. The constructor also needs a string to be used in the logging output. It's probably a good idea to use a string that matches the name of the instance. So, for example, the error instance gets an `"error"` string.

It's these instances that get passed to the `log` function, like this:

```
MereMemo::log(error) << message;
```

One thing to note is the namespace of the `LogLevel` instances. Because we're testing the logging library itself, we'll be calling `log` from within tests. Each test body is actually part of the `run` method of a test class defined with one of the `TEST` macros. The test class itself is in an unnamed namespace. I wanted to avoid needing to specify the `MereMemo` namespace when using one of the log levels, like this:

```
MereMemo::log(MereMemo::error) << message;
```

It's much easier to type just `error` instead of `MereMemo::error`. So, the solution (for now) is to declare the log-level instances in the global namespace inside `LogTags.h`. I recommend that when you define your own tags for your project, you declare the tags in your project's namespace. Something like this would work:

```
#ifndef YOUR_PROJECT_LOGTAGS_H
#define YOUR_PROJECT_LOGTAGS_H

#include <MereMemo/Log.h>

namespace yourproject
{

inline MereMemo::LogLevel error("error");
inline MereMemo::LogLevel info("info");
inline MereMemo::LogLevel debug("debug");

} // namespace yourproject

#endif // YOUR_PROJECT_LOGTAGS_H
```

Then, when you are writing your code in your project that is part of your own namespace, you can refer to the tags such as `error` directly without needing to specify a namespace. You can use whichever namespace you want in place of `yourproject`. You can see a good example of a project that uses both the logging library and the testing library in *Chapter 14, How To Test Services.*

Also, note that you should refer to the `Log.h` file from within your own project as a separate project and use angle brackets. This is just like what we did when we started work on the logging library and had to start referring to the unit test library include with angle brackets.

One extra benefit of passing an instance of `MereMemo::LogLevel` to the `log` function is that we no longer need to specify the namespace to the `log` function. The compiler knows to look in namespaces that function arguments use when trying to resolve the function name. The simple act of passing `error` to the `log` function lets the compiler figure out that the `log` function is the one defined in the same namespace as the `error` instance. I actually thought about this benefit once the code was working and I could try to call `log` without a namespace. I was then able to add a test to `Tags.cpp` that looks like this:

```
TEST("log needs no namespace when used with LogLevel")
{
    log(error) << "no namespace";
}
```

Here, you can see that we can call `log` directly without specifying the `MereMemo` namespace, and we can do this because the compiler knows that the `error` instance being passed is itself a member of `MereMemo`.

If we try to call `log` without any arguments, then we'll need to go back to using `MereMemo::log` instead of just `log`.

Also, notice how this new test was identified. It's an alternate usage that simplifies the code, and making a test helps make sure that we don't do anything later that will break the simpler syntax. The new test has no confirmations either. That's because the test exists just to make sure that the call to `log` without the namespace compiles. We already know that `log` can send a log message to the log file because the other test confirms this. This test doesn't need to duplicate the confirmation. If it compiles, then it has done its job.

The only thing we need now is a definition of the `LogLevel` class. Remember that we really want a general-purpose tagging solution and that a log level should be just one type of tag. There shouldn't be anything special about a log level versus any other tag. We might as well define a `Tag` class too and make `LogLevel` inherit from `Tag`. Put the two new classes at the beginning of `Log.h` just inside the `MereMemo` namespace, like this:

```
class Tag
{
public:
    virtual ~Tag () = default;
```

```cpp
        std::string key () const
        {
            return mKey;
        }

        std::string text () const
        {
            return mText;
        }

    protected:
        Tag (std::string const & key, std::string const & value)
        : mKey(key), mText(key + "=\"" + value + "\"")
        { }

    private:
        std::string mKey;
        std::string const mText;
    };

    class LogLevel : public Tag
    {
    public:
        LogLevel (std::string const & text)
        : Tag("log_level", text)
        { }
    };
```

Make sure both classes are defined inside the MereMemo namespace. Let's start with the Tag class, which should not be used directly. The Tag class should be a base class so that a derived class can specify the key to be used. The purpose of the Tag class is really just to make sure that the text output follows the key="value" format.

The LogLevel class inherits from the Tag class and only requires the text of the log level. The key is hardcoded to always be log_level, which enforces consistency. We get the consistency of the values when we declare instances of LogLevel with specific strings and then use the defined instances when calling log.

The logging library supports tags and even log-level tags but doesn't define any specific log levels itself. The library also doesn't try to order the log levels so that something such as `error` is a higher or lower level than `debug`. Everything is just a tag consisting of a key and a value.

Now that we have the `LogLevel` and `Tag` classes, how are they used by the `log` function? We'll first need a new overload of `log` that accepts a `Tag` parameter, like this:

```
inline std::fstream log (Tag const & tag)
{
    return log(to_string(tag));
}
```

Place this new `log` function right after the existing `log` function and still inside the `MereMemo` namespace in `Log.h`. The new `log` function will convert the tag into a string and pass the string to the existing `log` function. We'll need to define a `to_string` function that can be placed right after the definition of the `Tag` class, like this:

```
inline std::string to_string (Tag const & tag)
{
    return tag.text();
}
```

The `to_string` function just calls the `text` method in the `Tag` class to get the string. Do we really need a function for this? Couldn't we just call the text method directly from within the new overloaded `log` function? Yes, we could have, but it's a common practice in C++ to provide a function called `to_string` that knows how to convert a class into a string.

All these new functions need to be declared inline because we're going to keep the logging library as a single include file that another project can simply include to begin logging. We want to avoid declaring functions in the `Log.h` file and then implementing them inside a `Log.cpp` file because that would require users to add `Log.cpp` to their project, or it would require that the logging library be built as a library and then linked into the project. By keeping everything in a single header file, we make it easier for other projects to use the logging library. It's not really a library—it's just a header file that gets included. We'll still refer to it as the logging library, though.

The existing `log` function needs to be modified to accept a string. It actually used to accept a string for the message to be logged until we removed that and returned a stream instead that the caller uses to specify the message along with any other information to be logged. We're going to put a string parameter back in the `log` function and call it `preMessage`. The `log` function will still return a stream that the caller can use. The `preMessage` parameter will be used to pass the formatted tag, and the `log` function will output the `preMessage` before returning the stream for the caller to use for the other information to be logged. The modified `log` function looks like this:

```cpp
inline std::fstream log (std::string_view preMessage = "")
{
    auto const now = std::chrono::system_clock::now();
    std::time_t const tmNow =
        std::chrono::system_clock::to_time_t(now);
    auto const ms = duration_cast<std::chrono::milliseconds>(
        now.time_since_epoch()) % 1000;

    std::fstream logFile("application.log", std::ios::app);
    logFile << std::endl
        << std::put_time(std::gmtime(&tmNow),
            "%Y-%m-%dT%H:%M:%S.")
        << std::setw(3) << std::setfill('0')
        << std::to_string(ms.count())
        << " " << preMessage << " ";

    return logFile;
}
```

The preMessage parameter has a default value so that the log function can still be called without a log-level tag. All the log function does is send a timestamp, then the preMessage parameter to the stream, followed by a single space, before letting the caller have access to the returned stream.

Note that we still want the log-level tag to be separated from the timestamp with a space too. If there is no log level specified, then the output will have two spaces, which is a detail that will be fixed soon.

We have everything we need now to log with a log level that the new tests make use of:

```cpp
MereMemo::log(error) << message;
```

And building and running the project shows everything passes:

```
Running 1 test suites
-------------- Suite: Single Tests
------- Test: Message can be tagged in log
Passed
------- Test: log needs no namespace when used with LogLevel
Passed
------- Test: Simple message can be logged
Passed
```

```
------- Test: Complicated message can be logged
Passed
---------------------------------
Tests passed: 4
Tests failed: 0
```

And looking at a new log file shows the log levels, as expected:

```
2022-06-25T23:52:05.842 log_level="error" simple tag 7529
2022-06-25T23:52:05.844 log_level="error" no namespace
2022-06-25T23:52:05.844  simple 248 with more text.
2022-06-25T23:52:05.844  complicated 637 double=3.14 quoted="in
quotes"
```

The first two entries make use of the new log level. The second one is the one we only wanted to make sure compiles. The third and fourth logs are missing a log level. That's because they never specified a log level. We should fix this and enable some tags to have default values, which would let us add log levels without specifying a log level so that we can make sure that every log message entry has a log level. The third and fourth entries also have an extra space, which will be fixed too. The next section will add the ability to specify default tags.

Before moving on, notice one more thing. The complicated log entry actually looks like it uses tags already. That's because we formatted the message with a key="value" format. It's common to include quotation marks around text values and to not use quotation marks around numbers. The quotation marks help define the entire value when the text has spaces inside of the value, while numbers don't need spaces and therefore don't need quotation marks.

Also, note that we don't add spaces around the equals sign. We don't log the following:

```
double = 3.14
```

The reason we don't log this is that the extra spaces are not needed and only make it harder to process the log output. It might be easier to read with spaces, but trying to automate the processing of log files with scripts is harder with spaces.

Likewise, we don't use commas between tags. So, we don't do this:

```
double=3.14, quoted="in quotes"
```

Adding commas between tags might make it easier to read, but they are just one more thing that must be handled by code that needs to programmatically process log files. Commas are not needed, so we won't be using them.

Now, we can proceed to add default tags.

Adding default tag values

The previous section identified the need to sometimes add a tag to log messages even if the tag is not given to the log function. We can use this to add a default log-level tag or any other default value needed for any tag.

With this feature, we're starting to get to the need for the logging library to support configuration. What I mean is that we want to be able to tell the logging library how to behave before we call log, and we want the logging library to remember this behavior.

Most applications support logging only after the configuration is set once at the beginning of the application. This configuration setup is usually done at the beginning of the main function. So, let's focus on adding some simple configuration that will let us set some default tags and then use those default tags when logging. If we encounter both a default tag and a tag with the same key used during a call to the log function, then we will use the tag provided in the call to log. In other words, the default tags will be used unless overridden in the call to log.

We'll start with what it will take to set default tag values. This is a case where we won't actually have a test for setting a default value inside of main, but we will have a test to make sure that a default value set in main does appear in the log output from within a test. And we might as well design the solution so that default values can be set at any time and not just from inside the main function. This will let us test the setting of a default value directly instead of relying on main.

Even though the following code isn't inside of a test, we can still modify main first to make sure that the solution is something we like. Let's change main to look like this:

```
#include "../Log.h"

#include "LogTags.h"

#include <MereTDD/Test.h>

#include <iostream>

int main ()
{
    MereMemo::addDefaultTag(info);
    MereMemo::addDefaultTag(green);

    return MereTDD::runTests(std::cout);
}
```

We'll include `Log.h` so that we can get a definition of a new `addDefaultTag` function that we'll write, and we'll include `LogTags.h` to get access to the `info` log level and a new tag for a color. Why a color? Because when adding new tests, we want to look for simple and general use cases. We already have the `LogLevel` tag defined by the logging library, and the only thing we need to do is define specific instances with their own values. But we haven't yet defined our own tags, and this seems like a good place to check that custom tags work too. The usage flows well, and it seems reasonable that users would want to define multiple default tags.

It's easy to go too far and add a bunch of new functionality that needs to be tested, but adding related scenarios such as the two default tags `info` and `green` that serve to make a test more generic is okay. At least, it's the type of thing I would do in one step. You might want to make these two separate tests. I figure that we can add a single test that just makes sure both tags are present even if not provided to the `log` function. The fact that one tag type is provided by the logging library and the other is custom is not enough for me to require separate tests. I'll be happy if they both appear in the log output.

Let's add a test now to `Tags.cpp` that looks like this:

```
TEST("Default tags set in main appear in log")
{
    std::string message = "default tag ";
    message += Util::randomString();
    MereMemo::log() << message;

    std::string logLevelTag = " log_level=\"info\" ";
    std::string colorTag = " color=\"green\" ";
    bool result = Util::isTextInFile(message,
        "application.log",
        {logLevelTag, colorTag});
    CONFIRM_TRUE(result);
}
```

As it turns out, I'm glad that I did add two default tags instead of just one because when writing the test, I started thinking about how to verify they both appear in the log file, and that's when I realized that the `isTextInFile` function is too rigid for what we now need. The `isTextInFile` function worked okay when we were only interested in checking if a specific string appeared in a file, but we're working with tags now, and the order that the tags appear in the output is not specified. The important part is that we can't reliably create a single string that will always match the order of the tags in the output, and we definitely don't want to start checking for all possible tag orders.

What we want is the ability to first identify a specific line in the output. This is important because we might have many log file entries that have the same log level or the same color, but the message with the random number is more specific. Once we find a single line in the file that matches the random number, what we really want is to check that same line to make sure all the tags are present. The order within the line is not important.

So, I changed the `isTextInFile` function to take a third parameter, which will be a collection of strings. Each of these strings will be a single tag value to check. This actually makes the test easier to understand. We can leave the message unchanged and use it as the first argument to identify the line we want to find within the log file. Assuming we find that line, we then pass individually formatted tags in the `key="value"` format as a collection of strings to verify that they each exist in the same line found already.

Notice also that the tag strings begin and end with a single space. This makes sure that the tags are separated properly with spaces and that we also don't have any commas at the end of a tag value.

We should fix the other test that checks for the existence of the log level, like this:

```
TEST("Message can be tagged in log")
{
    std::string message = "simple tag ";
    message += Util::randomString();
    MereMemo::log(error) << message;

    std::string logLevelTag = " log_level=\"error\" ";
    bool result = Util::isTextInFile(message,
            "application.log",
        {logLevelTag});
    CONFIRM_TRUE(result);
}
```

We no longer need to append the message to the end of a formatted log-level tag. We just pass the single `logLevelTag` instance as the single value in the collection of additional strings to check. Now that we have default tag values set in `main`, there is no guarantee of the order of the tags. So, we could have failed this test because the color tag happened to come between the error tag and the message. All we check is that the message appears in the output and that the error tag also exists somewhere in the same log-line entry.

Let's enhance the `isTextInFile` function now to accept a vector of strings in a third parameter. The vector should have a default value of an empty collection in case the caller just wants to verify that a file contains some simple text without also looking for additional strings on the same line. And while we're doing this, let's add a fourth parameter, which will also be a vector of strings. The fourth parameter will check to make sure that its strings are *not* found in the line. The updated function declaration looks like this in `Util.h`:

```
#include <string>
```

```cpp
#include <string_view>
#include <vector>

struct Util
{
    static std::string randomString ();

    static bool isTextInFile (
        std::string_view text,
        std::string_view fileName,
        std::vector<std::string> const & wantedTags = {},
        std::vector<std::string> const & unwantedTags = {});
};
```

We need to include vector and make sure to give the extra parameters default empty values. The implementation in Util.cpp looks like this:

```cpp
bool Util::isTextInFile (
    std::string_view text,
    std::string_view fileName,
    std::vector<std::string> const & wantedTags,
    std::vector<std::string> const & unwantedTags)
{
    std::ifstream logfile(fileName.data());
    std::string line;
    while (getline(logfile, line))
    {
        if (line.find(text) != std::string::npos)
        {
            for (auto const & tag: wantedTags)
            {
                if (line.find(tag) == std::string::npos)
                {
                    return false;
                }
            }
            for (auto const & tag: unwantedTags)
```

```
            {
                if (line.find(tag) != std::string::npos)
                {
                    return false;
                }
            }
        }
        return true;
    }
    return false;
}
```

The change adds an extra `for` loop once we find the line identified by the `text` parameter. For all the wanted tags provided, we search through the line again to make sure each tag exists. If any of them are not found, then the function returns `false`. Assuming it finds all the tags, then the function returns `true`, just like before.

Almost the same thing happens for the unwanted tags except that the logic is reversed. If we find an unwanted tag, then the function returns `false`.

All we need now is to add the definition of the `Color` tag type and then the `green` color instance. We can add these to `LogTags.h`, like this:

```cpp
inline MereMemo::LogLevel error("error");
inline MereMemo::LogLevel info("info");
inline MereMemo::LogLevel debug("debug");

class Color : public MereMemo::Tag
{
public:
    Color (std::string const & text)
    : Tag("color", text)
    { }
};

inline Color red("red");
inline Color green("green");
inline Color blue("blue");
```

Building the project shows that I forgot to implement the addDefaultTag function that we started out using in main. Remember when I said that it's easy to get sidetracked? I started to add the function to Log.h, like this:

```
inline void addDefaultTag (Tag const & tag)
{
    static std::map<std::string, Tag const *> tags;

    tags[tag.key()] = &tag;
}
```

This is a great example of how writing the usage first helped with the implementation. What we need to do is store the tag passed to the addDefaultTag function so that it can be retrieved later and added to log messages. We first need a place to store the tags so that the function declares a static map.

Originally, I wanted the map to make a copy of the tag, but that would have required changing the Tag class so that it could be used directly instead of working with derived classes. I like how the derived classes help with keeping the key consistent and didn't want to change that part of the design.

So, instead, I decided that the collection of tags would use pointers. The problem with using pointers is that it's not obvious to the caller of addDefaultTag that the lifetimes of any tags passed to the function must remain valid for as long as the tag remains in the default tag collection.

We can still make copies and store the copies in unique pointers, but that requires either extra work for the caller of addDefaultTag or some method that knows how to clone a tag. I don't want to add extra complexity to the code in main that calls addDefaultTag and force that code to make a copy. We've already written the code in main, and we should strive to keep that code as-is because it was written using TDD principles and provides the solution we will be most happy with.

To avoid lifetime surprises, we should add a clone method to the Tag-derived classes. And because we are using a map in addDefaultTag and have identified the need for unique pointers, we need to include map and memory at the top of Log.h, like this:

```
#include <chrono>
#include <ctime>
#include <fstream>
#include <iomanip>
#include <iostream>
#include <map>
#include <memory>
#include <string>
#include <string_view>
```

Now, let's implement the correct addDefaultTag function to make a copy of the passed-in tag instead of storing a pointer directly to the caller's variable. This will free up the caller so that the tags passed in no longer have to remain alive indefinitely. Add this code to Log.h right after the LogLevel class:

```cpp
inline std::map<std::string, std::unique_ptr<Tag>> &
getDefaultTags ()
{
    static std::map<std::string, std::unique_ptr<Tag>> tags;
    return tags;
}


inline void addDefaultTag (Tag const & tag)
{
    auto & tags = getDefaultTags();
    tags[tag.key()] = tag.clone();
}
```

We use a helper function to store the collection of default tags. The collection is static, so it gets initialized to an empty map the first time the tags are requested.

We need to add a pure virtual clone method to the Tag class that will return a unique pointer. The method declaration can go right after the text method and looks like this:

```cpp
    std::string text () const
    {
        return mText;
    }


    virtual std::unique_ptr<Tag> clone () const = 0;

protected:
```

And now, we need to add the clone method implementation to both the LogLevel and Color classes. The first looks like this:

```cpp
class LogLevel : public Tag
{
public:
    LogLevel (std::string const & text)
    : Tag("log_level", text)
```

```
    { }

    std::unique_ptr<Tag> clone () const override
    {
        return std::unique_ptr<Tag>(
            new LogLevel(*this));
    }
};
```

And the implementation for the `Color` class looks almost identical:

```
class Color : public MereMemo::Tag
{
public:
    Color (std::string const & text)
    : Tag("color", text)
    { }

    std::unique_ptr<Tag> clone () const override
    {
        return std::unique_ptr<Tag>(
            new Color(*this));
    }
};
```

Even though the implementations look almost identical, each makes a new instance of the specific type involved, which gets returned as a unique pointer to `Tag`. This is the complexity that I was hoping to avoid when I started, but it's better to add complexity to the derived classes instead of placing extra and unexpected requirements on the caller of `addDefaultTag`.

We're now ready to build and run the test application. One of the tests fails, like this:

```
Running 1 test suites
-------------- Suite: Single Tests
------- Test: Message can be tagged in log
Passed
------- Test: log needs no namespace when used with LogLevel
Passed
------- Test: Default tags set in main appear in log
```

```
Failed confirm on line 37
    Expected: true
------- Test: Simple message can be logged
Passed
------- Test: Complicated message can be logged
Passed
-----------------------------------
Tests passed: 4
Tests failed: 1
```

The failure is actually a good thing and is part of the TDD process. We wrote the code as we intended it to be used in `main`, and wrote a test that would verify that the default tags appear in the output log file. The default tags are missing, and that's because we need to change the `log` function so that it will include the default tags.

Right now, the `log` function only includes the tags that are directly provided—or, I should say, the tag that is directly provided because we don't yet have a way to log multiple tags. We'll get there. One thing at a time.

Our `log` function currently has two overloaded versions. One takes a single `Tag` parameter and turns it into a string that it passes to the other. Once the tag is turned into a string, it becomes harder to detect which tags are currently being used, and we'll need to know that so that we don't end up logging both a default tag and a directly specified tag with the same key.

For example, we don't want a log message to include both `info` and `debug` log levels because the log was made with `debug` while `info` was the default. We only want the `debug` tag to appear because it should override the default.

We need to pass the tag to the `log` function that does the output as a `Tag` instance instead of a string. Instead of a single `Tag` instance, though, let's let callers pass more than one tag when calling `log`. Should we let the number of tags be unlimited? Probably not. Three seems like a good amount. If we need more than three, we'll come up with a different solution or add more.

I thought about different ways to write a `log` function that takes a variadic number of tags using templates. While it might be possible, the complexity quickly became unworkable. So, instead, here are three overloads of `log` that turn the parameters into a vector of `Tag` pointers:

```
inline auto log (Tag const & tag1)
{
    return log({&tag1});
}
```

```
inline auto log (Tag const & tag1,
    Tag const & tag2)
{
    return log({&tag1, &tag2});
}

inline auto log (Tag const & tag1,
    Tag const & tag2,
    Tag const & tag3)
{
    return log({&tag1, &tag2, &tag3});
}
```

These functions replace the earlier `log` function that converted the tag into a string. The new functions create a vector of `Tag` pointers. We might eventually need to call `clone` to create copies instead of using pointers to the caller's arguments, but for now, this works, and we don't have to worry about the lifetime issues we had with the default tags.

We'll need to include `vector` at the top of `Log.h`, and while implementing the `log` function that actually does the logging, I ended up needing `algorithm` too. The new include section looks like this:

```
#include <algorithm>
#include <chrono>
#include <ctime>
#include <fstream>
#include <iomanip>
#include <iostream>
#include <map>
#include <memory>
#include <string>
#include <string_view>
#include <vector>
```

And now, to the changes to the `log` function that does the logging. It looks like this:

```
inline std::fstream log (std::vector<Tag const *> tags = {})
{
    auto const now = std::chrono::system_clock::now();
    std::time_t const tmNow =
```

```
        std::chrono::system_clock::to_time_t(now);
    auto const ms = duration_cast<std::chrono::milliseconds>(
        now.time_since_epoch()) % 1000;

    std::fstream logFile("application.log", std::ios::app);
    logFile << std::endl
        << std::put_time(std::gmtime(&tmNow),
            "%Y-%m-%dT%H:%M:%S.")
        << std::setw(3) << std::setfill('0')
        << std::to_string(ms.count());

    for (auto const & defaultTag: getDefaultTags())
    {
        if (std::find_if(tags.begin(), tags.end(),
            [&defaultTag](auto const & tag)
            {
                return defaultTag.first == tag->key();
            }) == tags.end())
        {
            logFile << " " << defaultTag.second->text();
        }
    }
    for (auto const & tag: tags)
    {
        logFile << " " << tag->text();
    }
    logFile << " ";

    return logFile;
}
```

Instead of accepting a string of pre-formatted tags, the function now takes a vector of Tag pointers with a default value of an empty collection. As far as this function is concerned, there can be an unlimited number of tags. The limit of three tags comes only because of the overloaded log functions that take up to three tags.

The default value for the tags vector lets callers continue to be able to call log with no arguments.

The first part of the function that formats the timestamp, opens the log file, and prints the timestamp remains unchanged, except that we no longer display a pre-formatted string for the tags.

The changes start with the first `for` loop, which looks at each default tag. We want to try finding the same tag key in the vector of tag pointers. If we find the same key, then we skip the default tag and try the next one. If we don't find the same key, then we display the default tag.

To do the searching, we use the `std::find_if` algorithm and provide a lambda that knows how to compare the keys.

After displaying only the default tags that were not overridden, the code goes through a second `for` loop to display all the tags passed in directly.

Building and running the test application shows that all the tests pass, and the log file now contains default tags for all the entries, like this:

```
2022-06-26T06:24:26.607 color="green" log_level="error" simple
tag 4718
2022-06-26T06:24:26.609 color="green" log_level="error" no
namespace
2022-06-26T06:24:26.609 color="green" log_level="info" default
tag 8444
2022-06-26T06:24:26.609 color="green" log_level="info" simple
4281 with more text.
2022-06-26T06:24:26.610 color="green" log_level="info"
complicated 8368 double=3.14 quoted="in quotes"
```

All the log messages contain the color tag set to `"green"`, and they all contain the `log_level` tag with either the default value of `"info"` or the overridden value of `"error"`. For the test that overrides the default value, let's make sure that the default value does not exist. We can make use of the unwanted tags parameter in the `isTextInFile` function, like this:

```
TEST("Message can be tagged in log")
{
    std::string message = "simple tag ";
    message += Util::randomString();
    MereMemo::log(error) << message;

    // Confirm that the error tag value exists and that the
    // default info tag value does not.
    std::string logLevelTag = " log_level=\"error\" ";
    std::string defaultLogLevelTag = " log_level=\"info\" ";
```

```
        bool result = Util::isTextInFile(message,
            "application.log",
            {logLevelTag}, {defaultLogLevelTag});
        CONFIRM_TRUE(result);
}
```

Should the extra check that the default tag value does not exist in the log file be added to a separate test? The benefit of a separate test is that it makes it clear what is being tested. The downside is that the test will be almost identical to this one. It's something to think about. In this case, I think the extra check and comment in the existing test is enough.

Before moving on, we need to add a test for the feature that I slipped in for multiple tags. I really should have written a test for this first before enhancing the code to support multiple tags, but for the purposes of explaining the code, it was much more direct to just explain the idea of multiple tags once instead of going back and adding the extra explanation.

Let's quickly add a new type of Tag called Size with a few named instances in LogTags.h, like this:

```
class Size : public MereMemo::Tag
{
public:
    Size (std::string const & text)
    : Tag("size", text)
    { }

    std::unique_ptr<Tag> clone () const override
    {
        return std::unique_ptr<Tag>(
            new Size(*this));
    }
};

inline Size small("small");
inline Size medium("medium");
inline Size large("large");
```

And now, here is a test for multiple tags:

```
TEST("Multiple tags can be used in log")
{
```

```
std::string message = "multi tags ";
message += Util::randomString();
MereMemo::log(debug, red, large) << message;

std::string logLevelTag = " log_level=\"debug\" ";
std::string colorTag = " color=\"red\" ";
std::string sizeTag = " size=\"large\" ";
bool result = Util::isTextInFile(message,
        "application.log",
      {logLevelTag, colorTag, sizeTag});
CONFIRM_TRUE(result);
}
```

The log file contains the entry with all three tags, like this:

```
2022-06-26T07:09:31.192 log_level="debug" color="red"
size="large" multi tags 9863
```

We have the ability to log with up to three directly specified tags and multiple default tags. We need to eventually use the tags for more than just displaying information in the log file. We want to be able to filter log messages based on the tag values to control which log messages make it all the way to the log file and which are ignored. We're not quite ready for filtering. The next section will explore filtering options based on the tag values.

Exploring filtering options

Filtering log messages lets us write code that includes calls to log information at important places within the code but then ignore some of those logging calls. Why would we go to all the trouble of adding code to do logging but then not do the logging?

For some events in the code such as an error that gets detected, it makes sense to always log that event. Other places might be equally important even if they are not errors. Usually, these are places in the code where something gets created or deleted. I'm not talking about creating or deleting an instance of a local variable. I mean something major, such as the creation of a new customer account, the completion of a quest in an adventure game, or the deletion of an old data file to free up space. All of these are good examples of important events that should probably always be logged.

Other events might help a developer understand what a program was doing right before it crashed. These log messages act like signposts along a journey. They're not as big as the errors or major events, but they can let us figure out what a program was doing. These are usually good to log too because, without them, it can be hard to fix bugs. Sure—the error log might show clearly that something bad happened, but understanding what led up to the problem can be difficult without the signpost messages.

And sometimes, when we know the general idea of what led to a problem, we need even more details. This is where we sometimes want to turn off the logging because log messages such as these can sometimes be extremely verbose and cause the size of the log files to increase. They can also make it hard to see the bigger picture. Have you ever tried walking someplace with your eyes intently focused on the ground at your feet? You can get all the details of every step but might find that you get lost. Looking up to see the general direction makes it hard to also notice a small rock that can cause you to trip.

When writing code, we want to put all these types of log messages into the code because adding extra logging messages later can be difficult, especially if the program is running at a remote customer location. So, we want the code to try to log everything. And then, at runtime, we want to control exactly how much information appears in the log files. Filtering lets us control how much logging we see by ignoring some logging requests.

We're going to filter log messages based on tags and their values, but we have a problem.

Let's say that we want to ignore a log message unless it has a certain tagged value. The way our `log` function works now is that it immediately opens a log file and starts streaming a timestamp, then adds tags, and finally lets the caller send whatever else is needed.

The only way to know for certain if a log message should be allowed to complete is to look at the tags once they have been finalized. In other words, we need to let everything be sent as if it will be logged but without actually doing anything. Once we have the complete message, we can look at the message to see if it meets the criteria to be sent to the output file.

This means we need to do two things differently. First, we need to stop writing to a log file right away and collect everything in case we do eventually need to write it. And second, we need to know when a log message is complete. We can't simply return an open stream to the caller and let the caller do whatever they want with the stream. Or, I should say that we can't return a stream that directly modifies the output log file. Letting the caller work directly with the final output log file gives us no way to know when the caller is done so that we can finish up and either ignore the log or let it continue.

I know of three ways to determine when a potential log message is complete. The first is to put everything into a single function call. The function can accept a variable number of arguments, so we won't be limited to a fixed number. But because the entire log message is bundled into a single function call, we will know when we have everything. It might look like this:

```
MereMemo::log(info, " count=", 5, " with text");
```

I'm using a tag instance, a couple of string literals, and an integer number in this example. The string literals could instead be string variables or maybe function calls that return information to be logged. One of the string literals, together with the number, actually forms a `key=value` tag. The point is that the `log` function would know for certain exactly how much information was sent to be logged and we would know all the values. We could easily test the log message to see if it should be allowed to continue or if it should be ignored.

We even have the beginning of a solution like this already because we accept up to three tag instances in the `log` function.

The second way to determine when a log is complete is to use something to terminate the stream we have now. It might look like this:

```
MereMemo::log(info) << "count=" << 5 << " with text" <<
MereMemo::endlog;
```

Notice that we don't need the extra space inside the `"count="` string literal because the `log` function adds one for us after all the tags.

Or, we could even allow tags to be sent to the stream, like this:

```
MereMemo::log() << info << " count=" << 5 << " with text" <<
MereMemo::endlog;
```

And we're back to needing the leading space before the `count` string literal again. This is common for streams where the caller needs to manage spaces between streamed elements. The only place where we don't need to add a space is at the very first item streamed after the `log` function.

The main idea with the stream approach is that we need something at the end to let the logging library know that all the information is ready to be tested against the criteria to see if the log should be ignored or not.

I like the stream approach better. It feels more open to me—almost more natural. And because of operator precedence and the chaining of the streaming operators, we know the order in which the log line will be evaluated. That might not be very important, but it plays into the feeling that I like the streaming approach better.

With this second approach, the stream that the caller gets back from the `log` function can't be a `std::fstream` instance that is directly tied to the log file. Working with `fstream` directly would not let us ignore the log message because the information would already be sent to the file. Maybe we could return a stream that's tied to a string instead and let the terminating `endlog` element send the string that gets built to the log file or ignore it.

What happens if the terminating `endlog` element is forgotten? The terminating `endlog` element needs to evaluate the log and move it forward or ignore it. If `endlog` is forgotten, then the log message will not complete. The developer might not notice the problem until a need to look at the log file shows that the expected log message is always ignored.

The third approach is similar to the second but without the need for a terminating element that can be forgotten. Anytime a design relies on a person to remember to do something, there will almost certainly be cases where the required part is left out. By removing the need to remember to add a terminating marker, we get a better design that can no longer be misused due to a simple oversight.

We already know that we can't just return a stream tied directly to a log file. The third approach takes this a step further and returns a custom stream. We don't use a standard stream at all because we need to add code in the stream destructor that finishes the logging and decides to either let the message complete or ignore it.

This approach relies on specific object lifetime rules defined by C++. We need to know exactly when the destructor will run because we need the destructor to fill the role of a terminating `endlog` element. Other programming languages that use garbage collection to clean up deleted objects would not be able to support this third solution because the stream would not be deleted until some unspecified time in the future. C++ is very clear about when object instances get deleted, and we can rely on the order. For instance, we could make a call to `log` like this:

```
MereMemo::log(info) << "count=" << 5 << " with text";
```

The custom stream that `log` returns will be destructed at the semicolon, which ends the expression. The programmer cannot forget anything, and the stream will be able to run the same code that an explicit `endlog` element would trigger.

Maybe we could combine the best aspects of all three approaches. The first function call approach doesn't need a terminating element because it knows exactly how many arguments are being passed. The second terminating `endlog` approach is more open and natural and can work with a standard stream to a string, and the custom stream approach is open and natural too and avoids misuse.

I initially wanted to create a logging library that would be able to filter messages based on the entire message. While filtering on anything in the message seems to be the most flexible and powerful solution, it's also the most complicated to implement. We don't want to choose one design over another just because one is easier to code. We should choose a design based on the end usage that we will be happy with and find natural to use. Sometimes, complex implementations are a sign that the end use will also be complicated. A solution that might be less powerful overall but is easier to use will be better, as long as we don't take away anything that is required.

One filtering complexity that we should be able to remove without affecting the end use is to only look at tags formed through the `Tag`-derived classes. We should be able to drop the ability to filter a log message based on the content of manually crafted tags.

Another simplification we can make will be to only filter tags passed to the `log` function. This will combine the aspect of the first approach where the `log` function accepts multiple arguments with the custom streaming approach, which accepts an intuitive series of information in chunks. So, take a look at the following streaming example:

```
MereMemo::log(info) << green << " count=" << 5 << " with text";
```

Here, there are a total of three `key=value` tags. The first is the `info` tag, then the `green` tag, and then a tag formed manually with the count text and number. Instead of trying to filter based on all three tags, the only information we'll use for filtering will be the `info` tag because that is the only tag passed to the log function directly. We should also filter based on default tags because the log function knows about the default tags too. This makes it easy to understand what the log function does. The log function starts the logging and determines if anything that comes after it will be accepted or ignored.

If we want to consider the green tag in the filtering, then we just need to add it to the log function too, like this:

```
MereMemo::log(info, green) << "count=" << 5 << " with text";
```

This is the type of use that needs to be thought through with TDD. The result is not always the most powerful. Instead, the goal is to meet the needs of the user and be easy and intuitive to understand.

Because tags are becoming more important to this design, we should enhance them to support more than just text values. The next section will add new types of tags.

Adding new tag types

Since we're starting to refer to tags with numbers instead of text for the value, now would be a good time to add support for numeric and Boolean tags that don't need quotation marks around the value.

We're going to get slightly ahead of ourselves here and add some code that we don't have a test for. That's only because the additional support for numeric and Boolean tags is so similar to what we already have. This change is in Log.h in the Tag class. We need to add four extra constructors like this after the existing constructor that accepts a string:

```
protected:
    Tag (std::string const & key, std::string const & value)
    : mKey(key), mText(key + "=\"" + value + "\"")
    { }

    Tag (std::string const & key, int value)
    : mKey(key), mText(key + "=" + std::to_string(value))
    { }

    Tag (std::string const & key, long long value)
    : mKey(key), mText(key + "=" + std::to_string(value))
    { }

    Tag (std::string const & key, double value)
    : mKey(key), mText(key + "=" + std::to_string(value))
```

```
    { }

    Tag (std::string const & key, bool value)
    : mKey(key), mText(key + "=" + (value?"true":"false"))
    { }
```

Each constructor forms text following either the key="value" or the key=value syntax. To test the new constructors, we're going to need some new derived tag classes. All of these classes can go in LogTags.h. The two integral classes look like this:

```
class Count : public MereMemo::Tag
{
public:
    Count (int value)
    : Tag("count", value)
    { }

    std::unique_ptr<Tag> clone () const override
    {
        return std::unique_ptr<Tag>(
            new Count(*this));
    }
};

class Identity : public MereMemo::Tag
{
public:
    Identity (long long value)
    : Tag("id", value)
    { }

    std::unique_ptr<Tag> clone () const override
    {
        return std::unique_ptr<Tag>(
            new Identity(*this));
    }
};
```

We're not going to provide named instances of these tags. The earlier `Color` and `Size` tag types both have reasonable and common choices that make sense, but even they can be used directly if a strange color or an uncommon size needs to be logged. The new tags have no such common values.

Continuing, the double tag looks like this:

```
class Scale : public MereMemo::Tag
{
public:
    Scale (double value)
    : Tag("scale", value)
    { }

    std::unique_ptr<Tag> clone () const override
    {
        return std::unique_ptr<Tag>(
            new Scale(*this));
    }
};
```

And again, it has no obvious default values. Maybe we could provide a named value for 1.0 or some other specific values, but these seem like they would be best defined by the domain of the application. We're just testing a logging library and will go without named instances for this tag.

The Boolean tag looks like this:

```
class CacheHit : public MereMemo::Tag
{
public:
    CacheHit (bool value)
    : Tag("cache_hit", value)
    { }

    std::unique_ptr<Tag> clone () const override
    {
        return std::unique_ptr<Tag>(
            new CacheHit(*this));
    }
};
```

```
inline CacheHit cacheHit(true);
inline CacheHit cacheMiss(false);
```

And for this one, we have obvious named values for `true` and `false` that we can provide.

All of the new tag classes should give you an idea of what they can be used for. Many of these are very applicable for large financial microservices where, for example, values can take a long time to be calculated and need to be cached. Logging whether a result was due to a cache hit or miss is very valuable when figuring out the flow of a calculation.

We'd like to be able to pass one of the new tags to the stream returned by the `log` function, like this:

```
MereMemo::log(info) << Count(1) << " message";
```

To do so, we need to add a stream overload that knows how to handle the `Tag` class. Add this function to `Log.h` right after the `to_string` function:

```
inline std::fstream & operator << (std::fstream && stream, Tag
const & tag)
{
    stream << to_string(tag);
    return stream;
}
```

The function uses an *rvalue reference* to the stream because we're using the temporary stream returned from the `log` function.

Now, we can create a test that will log and confirm each of the new types. You could make separate tests for each type or put all of them into one test, like this:

```
TEST("Tags can be streamed to log")
{
    std::string messageBase = " 1 type ";
    std::string message = messageBase + Util::randomString();
    MereMemo::log(info) << Count(1) << message;

    std::string countTag = " count=1 ";
    bool result = Util::isTextInFile(message,
        "application.log", {countTag});
    CONFIRM_TRUE(result);
```

```
messageBase = " 2 type ";
message = messageBase + Util::randomString();
MereMemo::log(info) << Identity(123456789012345)
        << message;

std::string idTag = " id=123456789012345 ";
result = Util::isTextInFile(message, "application.log",
    {idTag});
CONFIRM_TRUE(result);

messageBase = " 3 type ";
message = messageBase + Util::randomString();
MereMemo::log(info) << Scale(1.5) << message;

std::string scaleTag = " scale=1.500000 ";
result = Util::isTextInFile(message, "application.log",
    {scaleTag});
CONFIRM_TRUE(result);

messageBase = " 4 type ";
message = messageBase + Util::randomString();
MereMemo::log(info) << cacheMiss << message;

std::string cacheTag = " cache_hit=false ";
result = Util::isTextInFile(message, "application.log",
    {cacheTag});
CONFIRM_TRUE(result);
}
```

The reason I wasn't so worried about creating this test before the code we added to enable the test is that we already thought through the desired usage before beginning.

The tag for double values might need some more work later to control the precision. You can see that it uses the default six decimals of precision. The log entries for the new test look like this:

```
2022-06-27T02:06:43.569 color="green" log_level="info" count=1
1 type 2807
```

```
2022-06-27T02:06:43.569 color="green" log_level="info"
id=123456789012345 2 type 7727
2022-06-27T02:06:43.570 color="green" log_level="info"
scale=1.500000 3 type 5495
2022-06-27T02:06:43.570 color="green" log_level="info" cache_
hit=false 4 type 3938
```

Notice how the message that is prepared for each log call is made unique with the numbers 1 to 4. This makes sure that in the rare case where a duplicate random number will be generated, none of the four log messages will have the same text.

We can now log default tags, tags provided directly to the log function, and tags that are streamed just like any other piece of information. Before we implement the actual filtering, there are some enhancements that the next section will make to improve the tag classes even more by reducing the amount of code that needs to be written for each tag class.

Refactoring the tag design with TDD

We have a base Tag class and several derived tag classes in the tests. Even though the logging library will only define the log-level tag, it should still make it easy for developers to create new derived tag classes. And right now, creating a new derived tag class is mostly boilerplate code that needs to be repeated over and over. We should be able to enhance the experience by using templates.

Here's what an existing derived tag class looks like:

```
class LogLevel : public Tag
{
public:
    LogLevel (std::string const & text)
    : Tag("log_level", text)
    { }

    std::unique_ptr<Tag> clone () const override
    {
        return std::unique_ptr<Tag>(
            new LogLevel(*this));
    }
};
```

The `LogLevel`-derived tag class is the only class like this that the logging library will provide. It defines the log-level tag without actually defining any specific log-level values. It's better to say that this class defines what a log level should be.

We can compare the `LogLevel` class to one of the other derived tag classes from the tests. Let's choose the `CacheHit` class, which looks like this:

```
class CacheHit : public MereMemo::Tag
{
public:
    CacheHit (bool value)
    : Tag("cache_hit", value)
    { }

    std::unique_ptr<Tag> clone () const override
    {
        return std::unique_ptr<Tag>(
            new CacheHit(*this));
    }
};
```

What can we improve about these classes? They're almost the same, with a few differences that can be moved into a template class. What's different about these two classes?

- The name, obviously. `LogLevel` versus `CacheHit`.

- The parent class namespace. `LogLevel` is already in the `MereMemo` namespace.

- The key string. `LogLevel` uses `"log_level"` while `CacheHit` uses `"cache_hit"`.

- The type of the value. `LogLevel` uses a `std::string` value while `CacheHit` uses a `bool` value.

That's all the differences. There should be no need to make developers recreate all this every time a new tag class is needed. And we're going to need to add more code to the tag classes in order to support filtering, so now is a great time to simplify the design.

We should be able to make the upcoming filtering changes without affecting any of the existing tests, but that will require design changes now. We're refactoring the design, and the tests will help make sure that the new design continues to behave just like the current design. The confidence we get from knowing everything still works is one of the benefits of using TDD.

The Tag class represents an interface that all tags support. We'll leave it as-is and simple. Instead of changing the Tag class, we'll introduce a new template class that can hold the clone method implementation and any upcoming filtering changes.

Change the LogLevel class in Log.h to use a new TagType template class that can use different types of values, like this:

```
template <typename T, typename ValueT>
class TagType : public Tag
{
public:
    std::unique_ptr<Tag> clone () const override
    {
        return std::unique_ptr<Tag>(
            new T(*static_cast<T const *>(this)));
    }


    ValueT value () const
    {
        return mValue;
    }
}

protected:
    TagType (ValueT const & value)
    : Tag(T::key, value), mValue(value)
    { }


    ValueT mValue;
};


class LogLevel : public TagType<LogLevel, std::string>
{
public:
    static constexpr char key[] = "log_level";


    LogLevel (std::string const & value)
```

```
        : TagType(value)
        { }
};
```

We still have a class called LogLevel that can be used just like before. It now specifies the type of the value, which is std::string, in the template argument to TagType, and the key string is now a constant array of chars that each derived tag class will define. The LogLevel class is simpler because it no longer needs to handle cloning.

The new TagType template class does most of the hard work. For right now, that work is just cloning, but we'll need to add more features in order to implement filtering. We should be able to put those upcoming features inside the TagType class and leave the derived tag classes unchanged.

The way this design works is based on something called the **Curiously Recurring Template Pattern (CRTP)** and involves a class inheriting from a template that is parameterized on the class itself. In this case, LogLevel inherits from TagType, and TagType is given LogLevel as one of its template parameters. This allows TagType to refer back to LogLevel from within the clone method in order to construct a new instance of LogLevel. Without the CRTP, then TagType would have no way to create a new LogLevel instance because it would not know what type to create.

And TagType needs to refer back to LogLevel one more time in order to get the name of the key. TagType does this again by referring to the type given to it by the CRTP in the T parameter.

The clone method is a little more complicated because when we're inside the clone method, we're in the TagType class, which means that the this pointer needs to be cast to the derived type.

We can now simplify the other derived tag types in LogTags.h. The Color and Size types both use std::string as the value type just like LogLevel, and they look like this:

```
class Color : public MereMemo::TagType<Color, std::string>
{
public:
    static constexpr char key[] = "color";

    Color (std::string const & value)
    : TagType(value)
    { }
};

class Size : public MereMemo::TagType<Size, std::string>
{
public:
```

```
    static constexpr char key[] = "size";

    Size (std::string const & value)
    : TagType(value)
    { }
};
```

The Count and Identity types both use integral value types of different lengths, and they look like this:

```
class Count : public MereMemo::TagType<Count, int>
{
public:
    static constexpr char key[] = "count";

    Count (int value)
    : TagType(value)
    { }
};

class Identity : public MereMemo::TagType<Identity, long long>
{
public:
    static constexpr char key[] = "id";

    Identity (long long value)
    : TagType(value)
    { }
};
```

The Scale type uses a double value type and looks like this:

```
class Scale : public MereMemo::TagType<Scale, double>
{
public:
    static constexpr char key[] = "scale";

    Scale (double value)
```

```
    : TagType(value)
    { }
};
```

And the `CacheHit` type uses a `bool` value type and looks like this:

```
class CacheHit : public MereMemo::TagType<CacheHit, bool>
{
public:
    static constexpr char key[] = "cache_hit";

    CacheHit (bool value)
    : TagType(value)
    { }
};
```

Each of the derived tag types is much simpler than before and can focus on what makes each one unique: the class name, the key name, and the type of the value.

The next section will create filtering tests based on logical criteria that will allow us to specify what should be logged, and we'll be using the simplified tag classes too with the `clone` method.

Designing tests to filter log messages

Filtering log messages is going to be one of the biggest features of the logging library. That's why this chapter is devoting so much effort to exploring ideas and enhancing the design. Most logging libraries offer some support for filtering, but usually, it's limited to just logging levels. And the logging levels are also usually ordered so that when you set one logging level, then you get all logs with a level equal to and either above or below the filtered level.

This always seemed arbitrary to me. Do the logging levels go up or down? Does setting the filtering level to `info` mean that you get `debug` too, or just `info` and `error` logs?

And this ignores the bigger problem of information overload. Once you do figure out how to get debug-level logs, they all get logged and the logs quickly fill up. I've even seen logs fill up so fast that the messages I was interested in were already zipped up and about to be deleted to save space before I could even exit the application to see what happened.

Our target customer for the logging library is a microservices developer. This means that the applications being worked on are probably large and distributed. Turning on debug logging everywhere, even within a single service, causes a lot of problems.

The logging library we're building will fix these problems, but we need to start simple. A test like this in Tags.cpp is a good start:

```
TEST("Tags can be used to filter messages")
{
    int id = MereMemo::createFilterClause();
    MereMemo::addFilterLiteral(id, error);

    std::string message = "filter ";
    message += Util::randomString();
    MereMemo::log(info) << message;

    bool result = Util::isTextInFile(message,
        "application.log");
    CONFIRM_FALSE(result);

    MereMemo::clearFilterClause(id);

    MereMemo::log(info) << message;

    bool result = Util::isTextInFile(message,
        "application.log");
    CONFIRM_TRUE(result);
}
```

The idea for this test is to first set a filter that will cause a log message to be ignored. We confirm that the message does not appear in the log file. Then, the test clears the filter and tries to log the same message again. This time, it should appear in the log file.

Normally, a filter match should allow a log to proceed, and no match should cause the message to be ignored. But when there are no filters set at all, then we should let everything through. Letting everything through without any filters lets users opt into filtering. If filtering is being used at all, then it controls the log output, but when there are no filters, then it would be strange to not let anything through. When the test sets a filter that does not match the log message, then the message does not appear in the log file because filtering has been enabled. When the filter is cleared, then we're assuming there are no other filters set, and all log messages will be allowed to proceed again.

We'll be filtering logs based on formulas in the **Disjunctive Normal Form** (**DNF**). The DNF specifies one or more clauses that are OR'ed together. Each clause contains literals that are AND'ed together. These are not literals in the C++ sense. Here, literal is a mathematical term. Each literal in a clause can either be AND'ed as-is or NOT'ed first. All this is Boolean logic and has the ability to represent any logical condition from simple to complex filters. Explaining all the details of DNF is not the purpose of this book, so I won't be explaining all the math behind DNF. Just know that DNF is powerful enough to represent any filter we can think of.

This is a case where a powerful solution is needed. Even so, we'll try to focus on the end use and keep the solution as easy to use as possible.

The test calls a `createFilterClause` function that returns an identifier for the clause created. Then, the test calls `addFilterLiteral` to add an `error` tag to the clause just created. What the test is trying to accomplish is to complete a log only if the `error` tag is present. If this tag is not present, then the log should be ignored. And remember that in order for a tag to be considered, it must either be present in the default tags or supplied directly to the `log` function.

Then, the test calls another function, `clearFilterClause`, which is intended to clear the filter clause just created and let everything be logged again.

Normally, the microservices developer won't run their application with filtering completely empty because that would allow all log messages to go through. Some amount of filtering will likely be in place at all times. As long as at least one filter clause is active, then filtering will only allow messages to proceed that match one of the clauses. By allowing multiple clauses, what we're doing is letting extra log messages go through because each additional clause has the chance to match more log messages. We'll have the ability to adjust what gets logged with a powerful system of Boolean logic.

A large project could then add tags that identify different components. The debug logs could be turned on only for certain components or for other matching criteria. The extra logic opens up more flexibility to increase the logging for interesting areas during debugging sessions while leaving other areas unaffected and logging at normal levels.

What happens if a tag is present in the default tags but gets overridden directly in the call to `log`? Should the default tag be ignored in favor of the explicit tag? I think so, and this will be a great test to include. Edge cases such as this really help define a project and improve the benefits gained by using TDD. Let's add the test now so that we don't forget. It looks like this:

```
TEST("Overridden default tag not used to filter messages")
{
    int id = MereMemo::createFilterClause();
    MereMemo::addFilterLiteral(id, info);

    std::string message = "override default ";
    message += Util::randomString();
    MereMemo::log(debug) << message;
```

```
    bool result = Util::isTextInFile(message,
        "application.log");
    CONFIRM_FALSE(result);


    MereMemo::clearFilterClause(id);
}
```

This test relies on the info tag already being set in the default tags. We should probably add the ability to test which tags are default so that the test can fail if info is not found in the default tags, and we need to make sure to clear the filter clause at the end of the test so that other tests are not affected. The previous test also cleared the clause but at a specific point in the test. Even so, the previous test should have a stronger guarantee that the test will not end with the filter clause still set. We should make use of a test teardown to always clear the filter clause at the end of any test that creates one.

Before continuing to add a teardown, the idea for the test that I started to explain is this. After setting a clause to only allow logs with the info tag, then the log message should have been allowed to continue because it will gain the info tag through the default set of tags. But instead, the log overrides the info tag with the debug tag. The end result is that the log message should not be found in the output log file.

To make sure that we always clear the filter clause even if a test fails and throws an exception before it reaches the end of the test, we need to define a setup and teardown class in Tags.cpp, like this:

```
class TempFilterClause
{
public:
    void setup ()
    {
        mId = MereMemo::createFilterClause();
    }


    void teardown ()
    {
        MereMemo::clearFilterClause(mId);
    }


    int id () const
    {
        return mId;
    }
```

```
private:
    int mId;
};
```

If you want more information about setup and teardown classes, refer to *Chapter 7, Test Setup and Teardown*.

It's okay for a test to clear the filters itself at the appropriate times. Adding an instance of SetupAndTeardown will make sure to call the clearFilterClause function even if it was already called. The first test from this section looks like this:

```
TEST("Tags can be used to filter messages")
{
    int id = MereMemo::createFilterClause();
    MereMemo::addFilterLiteral(id, error);

    std::string message = "filter ";
    message += Util::randomString();
    MereMemo::log(info) << message;

    bool result = Util::isTextInFile(message,
        "application.log");
    CONFIRM_FALSE(result);

    MereMemo::clearFilterClause(id);

    MereMemo::log(info) << message;

    result = Util::isTextInFile(message, "application.log");
    CONFIRM_TRUE(result);
}
```

The test now gets the clause ID from the setup and teardown instance. The ID is used to add the filter literal and to clear the filter clause at the correct time. The filter clause will be cleared again at the end of the test with no effect.

The second test from this section no longer needs to explicitly clear the filter itself and only needs to add the SetupAndTeardown instance, like this:

```
TEST("Overridden default tag not used to filter messages")
{
    MereTDD::SetupAndTeardown<TempFilterClause> filter;
    MereMemo::addFilterLiteral(filter.id(), info);

    std::string message = "override default ";
    message += Util::randomString();
    MereMemo::log(debug) << message;

    bool result = Util::isTextInFile(message,
        "application.log");
    CONFIRM_FALSE(result);
}
```

This test was calling clearFilterClause at the end to put the filters back in an unfiltered state. The test no longer needs to call clearFilterClause directly because relying on the SetupAndTeardown destructor is more reliable.

We have two filter tests that call functions that don't exist yet. Let's add the following function stubs to Log.h right after the addDefaultTag function:

```
inline int createFilterClause ()
{
    return 1;
}

inline void addFilterLiteral (int filterId,
    Tag const & tag,
    bool normal = true)
{
}

inline void clearFilterClause (int filterId)
{
}
```

The createFilterClause function just returns 1 for now. It will need to eventually return a different identifier for each clause created.

The `addFilterLiteral` function adds the given tag to the clause identified. The `normal` parameter will let us add literals that are NOT'ed or inverted by passing `false`. Be careful with the meaning of flags such as this. When I first wrote this, the flag was called `invert` and had a default value of `false`. I didn't notice the problem until writing a test for an inverted filter and it seemed strange to pass `true` in order to get an inverted literal. The test highlighted the backward usage while the initial function declaration let it slip by unnoticed.

And the `clearFilterClause` function does nothing for now. We'll need to have some sort of collection of clauses later that we can work with.

Stubbing out the filter functions lets us build and run the test application. We get two test failures, like this:

```
Running 1 test suites
-------------- Suite: Single Tests
------- Test: Message can be tagged in log
Passed
------- Test: log needs no namespace when used with LogLevel
Passed
------- Test: Default tags set in main appear in log
Passed
------- Test: Multiple tags can be used in log
Passed
------- Test: Tags can be streamed to log
Passed
------- Test: Tags can be used to filter messages
Failed confirm on line 123
    Expected: false
------- Test: Overridden default tag not used to filter
messages
Failed confirm on line 143
    Expected: false
------- Test: Simple message can be logged
Passed
------- Test: Complicated message can be logged
Passed
----------------------------------
Tests passed: 7
Tests failed: 2
```

The results are expected with TDD. We did the bare minimum needed to get the code building so that we can see the failures. We can add a little more implementation to the stubbed-out functions next.

I mentioned that we'll need a collection for the clauses. Add the following functions to Log.h, right before the stubbed-out filter functions:

```
struct FilterClause
{
    std::vector<std::unique_ptr<Tag>> normalLiterals;
    std::vector<std::unique_ptr<Tag>> invertedLiterals;
};

inline std::map<int, FilterClause> & getFilterClauses ()
{
    static std::map<int, FilterClause> clauses;
    return clauses;
}
```

The pattern is similar to what we did for the default tags. There is a function called getFilterClauses that returns a reference to a static map of FilterClause objects, and the FilterClause struct is defined to hold a couple of vectors for the normal and inverted literals. The literals are pointers to the tags that we get from cloning.

The createFilterClause function can be implemented to use the collection of clauses, like this:

```
inline int createFilterClause ()
{
    static int currentId = 0;
    ++currentId;
    auto & clauses = getFilterClauses();
    clauses[currentId] = FilterClause();

    return currentId;
}
```

This function keeps track of the current id in a static variable that gets incremented each time the function is called. The only other task that needs to be done is to create an empty filter clause record. The id is returned to the caller so that the filter clause can be modified or cleared later.

The addfilterLiteral function can be implemented like this:

```
inline void addFilterLiteral (int filterId,
    Tag const & tag,
    bool normal = true)
{
    auto & clauses = getFilterClauses();
    if (clauses.contains(filterId))
    {
        if (normal)
        {
            clauses[filterId].normalLiterals.push_back(
                tag.clone());
        }
        else
        {
            clauses[filterId].invertedLiterals.push_back(
                tag.clone());
        }
    }
}
```

This function makes sure that the clauses collection contains an entry for the given filter id before pushing back a cloned pointer to either the normal or inverted vector.

And the clearFilterClause function is the simplest because it just needs to get the collection and erase whichever filter clause exists with the given id like this:

```
inline void clearFilterClause (int filterId)
{
    auto & clauses = getFilterClauses();
    clauses.erase(filterId);
}
```

We still need to examine the filter clauses when logging, and that will be explained in the next section. When following TDD, it's good to get tests working to the point where the code builds and the tests fail when run. Let's get the tests to pass in the next section!

Controlling what gets logged

Earlier in this chapter when we were exploring filtering options, I mentioned that we will need a custom stream class instead of returning `std::fstream` from the `log` function. We need this so that we don't immediately send information to the log file. We need to avoid sending a log message directly to the log file because there could be filtering rules in place that could cause the log message to be ignored.

We also decided that we would make the decision to log or not based entirely on the default tags and any tags sent directly to the `log` function. We could have the `log` function make the decision and either return `std::fstream` if the log message should proceed or a fake stream if the log message should be ignored, but it's probably better to always return the same type. That seems like the simplest and most straightforward solution. Switching between stream types just seems like a more complicated solution that still requires a custom stream type.

And using a custom stream type will also let us fix a nagging problem where we have to put newlines *before* each log message instead of *after*. This has resulted in log files with an empty first line and the last line ending abruptly. We went with the temporary solution to put newlines before each log message because we didn't have anything at the time that would let us know when all the information had been streamed.

Well, a custom stream class will let us solve the nagging newline problem and give us a way to avoid writing log messages directly to the log file. Let's start with the new stream class. Create this class in `Log.h`, right before the `log` functions, like so:

```cpp
class LogStream : public std::fstream
{
public:
    LogStream (std::string const & filename,
        std::ios_base::openmode mode = ios_base::app)
    : std::fstream(filename, mode)
    { }

    LogStream (LogStream const & other) = delete;

    LogStream (LogStream && other)
    : std::fstream(std::move(other))
    { }

    ~LogStream ()
    {
```

```
            *this << std::endl;
    }


    LogStream & operator = (LogStream const & rhs) = delete;
    LogStream & operator = (LogStream && rhs) = delete;
};
```

We're going to fix one problem at a time. So, we'll continue to refactor this class until it does everything we need. Right now, it just inherits from `std::fstream`, so it won't solve the problem of writing directly to the log file. The constructor still opens the log file, and all the streaming capability is inherited from `fstream`.

What this class does solve is the newline problem. It solves this by sending `std::endl` to the stream in the class destructor. The constructor that opens the file based on the name provided and the destructor that adds the newline are really the only parts of this class that solve the problem. The rest of the class is needed to get the code to compile and work properly.

Because we added a destructor, that set off a chain reaction of other requirements. We now need to provide a copy constructor. We actually need the *move copy constructor* because streams tend to act strange when copied. Copying a stream is not a simple task, but moving a stream into another stream is much simpler and does everything we need anyway. We don't need to make any copies of the stream but we do need to return the stream from the `log` function, which means the stream either needs to be copied or moved. So, we explicitly delete the copy constructor and implement the move copy constructor.

We also delete both the assignment operator and the move assignment operator because we don't need to assign the stream either.

We can use the new `LogStream` class by modifying the `log` function to look like this:

```
inline LogStream log (std::vector<Tag const *> tags = {})
{
    auto const now = std::chrono::system_clock::now();
    std::time_t const tmNow =
        std::chrono::system_clock::to_time_t(now);
    auto const ms = duration_cast<std::chrono::milliseconds>(
        now.time_since_epoch()) % 1000;

    LogStream ls("application.log");
    ls << std::put_time(std::gmtime(&tmNow),
        "%Y-%m-%dT%H:%M:%S.")
        << std::setw(3) << std::setfill('0')
```

```
            << std::to_string(ms.count());

    for (auto const & defaultTag: getDefaultTags())
    {
        if (std::find_if(tags.begin(), tags.end(),
            [&defaultTag](auto const & tag)
            {
                return defaultTag.first == tag->key();
            }) == tags.end())
        {
            ls << " " << defaultTag.second->text();
        }
    }
    for (auto const & tag: tags)
    {
        ls << " " << tag->text();
    }
    ls << " ";

    return ls;
}
```

The log function now returns a LogStream instance instead of std::fstream. Inside the function, it creates a LogStream instance as if it were a fstream instance. The only thing that changes is the type. And we now have the file open mode defaulted to append, so we don't need to specify how to open the file. The name of the stream changed to ls because this is no longer a log file.

And then, when sending the initial timestamp, we no longer need to send an initial std::endl instance and can immediately start with the timestamp.

The only thing different when the test application runs after these changes is that the log file will no longer have an empty first line and all lines will end with a newline.

That's one small problem fixed. What about the bigger problem of writing directly to the log file? We still want to write to a standard stream because implementing our own stream class adds a lot of complexity we don't really need right now. So, instead of inheriting the LogStream class from std::fstream, we'll instead inherit from std::stringstream.

We're going to need to include sstream to get a definition of stringstream, and we might as well include ostream now too. We'll need ostream in order to change our streaming helper function in Log.h, which currently uses std::fstream, to look like this instead:

```
inline std::ostream & operator << (std::ostream && stream, Tag
const & tag)
{
    stream << to_string(tag);
    return stream;
}
```

We probably should have implemented this helper function to use `ostream` from the very beginning. This way, we can stream tags to any output stream. And because both `fstream` and `stringstream` are based on `ostream`, we can use this one helper function to stream to both.

Here are the updated includes for `Log.h`:

```
#include <algorithm>
#include <chrono>
#include <ctime>
#include <fstream>
#include <iomanip>
#include <map>
#include <memory>
#include <ostream>
#include <sstream>
#include <string>
#include <string_view>
#include <vector>
```

Technically, we don't need to include `ostream` because we get it already through including `fstream` and `stringstream`, but I like to include headers for things we are directly using. And while looking into the headers being included, I noticed that we were including `iostream`. I think I originally included `iostream` to get the definition of `std::endl`, but it seems that `endl` is actually declared in `ostream`. So, based on my rule to include headers being used, we should have been including `ostream` since the very beginning instead of `iostream`.

Back to `LogStream`, we need to change this class to inherit from `stringstream`, like this:

```
class LogStream : public std::stringstream
{
public:
    LogStream (std::string const & filename,
        std::ios_base::openmode mode = ios_base::app)
```

```
    : mProceed(true), mFile(filename, mode)
    { }


    LogStream (LogStream const & other) = delete;


    LogStream (LogStream && other)
    : std::stringstream(std::move(other)),
    mProceed(other.mProceed), mFile(std::move(other.mFile))
    { }


    ~LogStream ()
    {
        if (not mProceed)
        {
            return;
        }
        mFile << this->str();
        mFile << std::endl;
    }


    LogStream & operator = (LogStream const & rhs) = delete;
    LogStream & operator = (LogStream && rhs) = delete;


    void ignore ()
    {
        mProceed = false;
    }

private:
    bool mProceed;
    std::fstream mFile;
};
```

There is a new data member called mProceed that we set to true in the constructor. Since we no longer inherit from std::fstream, we now need a data member that is a file stream. We also need to initialize the mFile member. The move copy constructor needs to initialize the data members, and the destructor checks if the logging should proceed or not. If the logging should proceed, then the string content of `stringstream` is sent to the file stream.

We still haven't implemented the filtering, but we're getting close. This change gets us to the point where we can control the logging. The logging will proceed unless we call `ignore` before the destructor is run. This simple change will let us build and test to make sure that we haven't broken anything.

Running the test application shows the same two test failures as before related to the filtering. The main thing is that the other tests continue to pass, which shows that the changes to use `stringstream` are working as before when we were streaming directly to the file stream.

It's important when making pivotal changes such as switching streams to make sure that nothing breaks. This is why I made the change with a hardcoded choice to always log. We can use the TDD tests we already have to verify that the stream change works before making more changes to add filtering.

Let's take the next change to the `log` function in two parts. We're going to need the full collection of active tags after figuring out which of the default tags have been overridden. Instead of sending the tags to the stream, we can first put them into an active collection, like this:

```cpp
inline LogStream log (std::vector<Tag const *> tags = {})
{
    auto const now = std::chrono::system_clock::now();
    std::time_t const tmNow =
        std::chrono::system_clock::to_time_t(now);
    auto const ms = duration_cast<std::chrono::milliseconds>(
        now.time_since_epoch()) % 1000;

    LogStream ls("application.log");
    ls << std::put_time(std::gmtime(&tmNow),
        "%Y-%m-%dT%H:%M:%S.")
        << std::setw(3) << std::setfill('0')
        << std::to_string(ms.count());

    std::map<std::string, Tag const *> activeTags;
    for (auto const & defaultTag: getDefaultTags())
    {
        activeTags[defaultTag.first] = defaultTag.second.get();
    }
    for (auto const & tag: tags)
    {
        activeTags[tag->key()] = tag;
```

```
    }
    for (auto const & activeEntry: activeTags)
    {
        ls << " " << activeEntry.second->text();
    }
    ls << " ";


    // Filtering will go here.
    return ls;
}
```

This not only gets us the active collection, but it also seems simpler. We let the map handle the overrides by first putting all the default tags into the map and then putting all the provided tags into the map. Building and running the test application shows that the change hasn't broken anything new. So, we're ready for the next part, which is comparing the filtering clauses with the active tags.

The filtering needs to change the last part of the log function where the comment indicates that filtering will go, like this:

```
bool proceed = true;
for (auto const & clause: getFilterClauses())
{
    proceed = false;
    bool allLiteralsMatch = true;
    for (auto const & normal: clause.second.normalLiterals)
    {
        // We need to make sure that the tag is
        // present and with the correct value.
        if (not activeTags.contains(normal->key()))
        {
            allLiteralsMatch = false;
            break;
        }
        if (activeTags[normal->key()]->text() !=
            normal->text())
        {
            allLiteralsMatch = false;
            break;
```

```
            }
        }
        if (not allLiteralsMatch)
        {
            continue;
        }
        for (auto const & inverted:
            clause.second.invertedLiterals)
        {
            // We need to make sure that the tag is either
            // not present or has a mismatched value.
            if (activeTags.contains(inverted->key()))
            {
                if (activeTags[inverted->key()]->text() !=
                    inverted->text())
                {
                    break;
                }
                allLiteralsMatch = false;
                break;
            }
        }
        if (allLiteralsMatch)
        {
            proceed = true;
            break;
        }
    }

    if (not proceed)
    {
        ls.ignore();
    }
    return ls;
```

The logic is a little complicated, and this is a case where I found it easier to implement the logic almost completely instead of trying to break the change into multiple parts. Here's what the code does.

Because we are using DNF logic, we can treat each clause separately. We start out as if we will proceed with the log, just in case there are no filters set at all. If there are any filters, then for each one, we start out as if we will not proceed. But we also set a new `bool` variable that assumes that all the literals will match until proven otherwise. We'll treat a clause without any literals as a sign that we should proceed with the log.

For checking the literals, we have two types: normal and inverted. For the normal literals, the tags must all be present in the active tags and have matching values. If any tag is missing or has the wrong value, then we did not match all the literals for this clause. We'll continue because there might be another clause that will match. This is what I mean about treating each clause separately.

Assuming we matched all the normal literals, we must still check the inverted literals. Here, the logic is reversed, and we need to make sure that either the tag is not present or that it has the wrong value.

Once we've checked all the clauses or found a clause that matches all the literals, the code makes one last check to see if the log should proceed or not. If not, then we call `ignore`, which will stop the log message from being sent to the output log file.

This approach makes a decision about whether or not to proceed at the time that the `log` function is called based on the default tags and tags sent to the `log` function. We'll let the calling code send whatever information is desired to the stream. The information will only make it all the way to the output log file if `ignore` was not called.

Everything builds and runs now, and we get all tests passing again, like this:

```
Running 1 test suites
-------------- Suite: Single Tests
------- Test: Message can be tagged in log
Passed
------- Test: log needs no namespace when used with LogLevel
Passed
------- Test: Default tags set in main appear in log
Passed
------- Test: Multiple tags can be used in log
Passed
------- Test: Tags can be streamed to log
Passed
------- Test: Tags can be used to filter messages
Passed
------- Test: Overridden default tag not used to filter
messages
Passed
```

```
------- Test: Simple message can be logged
Passed
------- Test: Complicated message can be logged
Passed
----------------------------------
Tests passed: 9
Tests failed: 0
```

This shows that the filtering is working! At least, for the equality of the tags. Testing whether or not a tag is present with a matching value is a good start, but our microservices developer will need more power than this. Maybe we will need to log only if a count tag has a value greater than 100 or some other comparison that involves a numeric value greater or lesser than a specified filter value. This is what I meant when I said that I implemented the filtering logic *almost* completely. I got the logic and all the loops and breaks working for tag equality. We should be able to use the same basic structure of the code for relative comparisons in the next section.

There's one more thing to add before we start relative comparisons, and this is important. Whenever code is added like what I did with the DNF logic without a test to back it up, we need to add a test. Otherwise, a missed test has a way of getting pushed back until it's forgotten about completely.

And this new test helped out in another way. It caught a problem with the initial definition of the addFilterLiteral function. The original function defined a bool parameter called invert that had a default value of false. The default value meant that creating a normal literal could leave out the parameter and use the default. But to create an inverted literal, the function required that the true value be passed. This seemed backward to me. I realized that it would make more sense to pass false for this parameter to get an inverted literal, and that true should create a normal literal. So, I went back and changed the function definition and implementation. The test caught a problem with the usage of a function that went unnoticed at first.

Here is the new test that will create an inverted filter:

```
TEST("Inverted tag can be used to filter messages")
{
    MereTDD::SetupAndTeardown<TempFilterClause> filter;
    MereMemo::addFilterLiteral(filter.id(), green, false);

    std::string message = "inverted ";
    message += Util::randomString();
    MereMemo::log(info) << message;

    bool result = Util::isTextInFile(message,
```

```
                    "application.log");
        CONFIRM_FALSE(result);
}
```

Building and running shows that the new test passes, and we have confirmed that we can filter log messages that contain a matching tag when the filter is inverted. This test uses the default green tag that is added to log messages and makes sure that the log message does not appear in the output log file because of the presence of the green tag.

The next section will enhance the filtering to allow filtering based on relative values of tags instead of just exact matches.

Enhancing filtering for relative matches

TDD encourages incremental changes and enhancements when designing software. Write a test, get something working, and then write a more elaborate test that enhances the design. We've been following a TDD approach to designing the logging library, and the previous section was a great example. We got filtering working in the previous section, but only for tag equality.

In other words, we can now filter log messages based on the presence or absence of a tag that matches a filter literal tag. We compare the tags to see if both the key and value match. That was a great first step because even getting that far required a lot of work. Imagine if we had tried to go all the way and supported, for example, logging only if a count tag had a value greater than 100.

When designing software using TDD, it really helps to look for obvious steps that can be taken and confirmed to work before taking the next step. Some steps might be bigger than others, and that's okay as long as you don't go straight to the final implementation because that will only lead to longer development times and more frustration. It's so much better to confirm some parts of the design work as expected and have tests to make sure those parts continue to work. It's like building a house with a solid foundation. It's much better to make sure that the foundation really is solid before building the walls, and you want to have tests to make sure that the walls stay straight while the roof is being added.

We have working tests in place to make sure that basic filtering works. We are testing both normal and inverted literals. We check for matching tags by comparing the text of the tags, which works for all value types. For relative filters such as a count greater than 100, we're going to need a solution that knows how to compare the values with a numeric check instead of a string match.

We can start by figuring out how to represent a filter literal to check for greater or lesser numeric values. Here is a test that can go in Tags.cpp that sets a filter based on a count greater than 100:

```
TEST("Tag values can be used to filter messages")
{
    MereTDD::SetupAndTeardown<TempFilterClause> filter;
```

```
MereMemo::addFilterLiteral(filter.id(),
    Count(100, MereMemo::TagOperation::GreaterThan));

std::string message = "values ";
message += Util::randomString();
MereMemo::log(Count(1)) << message;

bool result = Util::isTextInFile(message,
    "application.log");
CONFIRM_FALSE(result);

MereMemo::log() << Count(101) << message;

result = Util::isTextInFile(message, "application.log");
CONFIRM_FALSE(result);

MereMemo::log(Count(101)) << message;

result = Util::isTextInFile(message, "application.log");
CONFIRM_TRUE(result);
}
```

What's new with this test? The main part is the way the Count tag is created. We previously only added a value when creating tags, like this:

```
Count(100)
```

And because we now need a way to specify if something should have a relative value, we need a place to say what kind of relative value and a method to communicate which relative value to use. I think an enumeration of various relative comparisons should work. We probably don't need more advanced relative comparisons such as "between" because we can always use DNF to express more elaborate comparisons. For a brief overview of how we're using DNF, refer to the *Designing tests to filter log messages* section of this chapter.

At the tag level, all we really need is to know how to compare one value against another. So, it makes sense to specify what type of comparison is needed when constructing a tag, like this:

```
Count(100, MereMemo::TagOperation::GreaterThan)
```

It might make sense to treat a tag with a comparison operator such as GreaterThan as a completely different type, but I think we can get away with a single type. With this solution, any tag can have a comparison operator, but it only makes sense to specify comparison operators for tags that will be used in filters.

What happens if a regular tag without a comparison operator is used in a filter? Then, we should treat that as an exact match because that's what the existing tests expect.

Back to the new test. It first creates a filter that should only let a message be logged if it has a count tag with a value greater than 100. It first tries to log a message with a count of only 1, and this is verified to not exist in the log file.

Then, the test creates a count of 101 but does not use the count tag directly in the log function call. This also should not make it to the output log file because we only want to filter on tags that are either default or directly specified when calling log.

And finally, the test calls log with a count tag of 101, and this message is verified to appear in the log file.

Now that we have a test, how will we get it working? Let's define comparison operations first in Log.h, right before the TagType class, like this:

```
enum class TagOperation
{
    None,
    Equal,
    LessThan,
    LessThanOrEqual,
    GreaterThan,
    GreaterThanOrEqual
};
```

We'll use the None operation for regular tags that only want to express a value. The Equal operation will act like the existing equality checks between tags. And the real change is to support less than, less than or equal, greater than, and greater than or equal comparisons.

We need to compare one tag with another without worrying about what the tags represent. A good way to do this is to declare a pure virtual method in the Tag class, just like what we did for cloning. The new method is called match and can go right after the clone method, like this:

```
virtual std::unique_ptr<Tag> clone () const = 0;

virtual bool match (Tag const & other) const = 0;
```

Here's where things get a bit difficult. I had wanted to wrap everything up in the TagType class. The idea was to first check the key of each tag being compared and make sure that the tags were the same. If they have the same key, then check the value. If they don't have the same key, then they must not match. At least, that was a good plan. I ran into problems when trying to implement the match method in one place that could compare strings with strings, numerics with numerics, and Booleans with Booleans. A tag such as CacheHit has a bool value type, and the only operation that makes sense is Equal comparisons. Tags based on strings need to compare differently than numerics. And if we really want to get elaborate, doubles should compare differently than an int type.

Each derived tag type could know how to compare, but I didn't want to change the derived types and make them each implement the match method, especially after all the work we went through to avoid the derived types implementing clone. The best solution I came up with was to create an additional set of intermediate classes that derive from TagType. Each new class is based on the type of the value. Since we only support five different tag value types, this is not a bad solution. The main benefit is that the derived tag types that the caller will use are only slightly affected. Here's a new StringTagType class that inherits from TagType so that you can see what I mean. Place this new class in Log.h, right after the TagType class:

```
template <typename T>
class StringTagType : public TagType<T, std::string>
{
protected:
    StringTagType (std::string const & value,
        TagOperation operation)
    : TagType<T, std::string>(value, operation)
    { }

    bool compareTagTypes (std::string const & value,
        TagOperation operation,
        std::string const & criteria) const override
    {
        int result = value.compare(criteria);
        switch (operation)
        {
        case TagOperation::Equal:
            return result == 0;

        case TagOperation::LessThan:
            return result == -1;
```

```
            case TagOperation::LessThanOrEqual:
                return result == 0 || result == -1;


            case TagOperation::GreaterThan:
                return result == 1;


            case TagOperation::GreaterThanOrEqual:
                return result == 0 || result == 1;


        default:
            return false;
        }
    }
};
```

This class is all about comparing string-based tags with other string-based tags. The class implements a new virtual method I'll explain in just a moment, called `compareTagTypes`. The only thing this method has to worry about is how to compare two strings based on an operation. One of the strings is called `value` and the other is called `criteria`. It's important to not mix up the `value` and `criteria` strings because, for example, while "ABC" is greater than "AAA", the same is not true the other way around. The method uses the `compare` method in the `std::string` class to do the comparisons.

You can see that the `StringTagType` class inherits from `TagType` and passes on the T derived type while hardcoding `std::string` for the value type. One interesting thing about the constructor is the need to repeat the template parameters when constructing `TagType` in the constructor initialization list. Normally, this should not be required, but maybe there is some arcane rule that only applies here that I'm not aware of whereby the compiler does not look at the `TagType` parameters in the parent class list to figure out the template parameters.

Before moving on to the changes in `TagType`, let's look at how a derived tag class such as `LogLevel` will use the new `StringTagType` intermediate class. Change the `LogLevel` class to look like this:

```
class LogLevel : public StringTagType<LogLevel>
{
public:
    static constexpr char key[] = "log_level";


    LogLevel (std::string const & value,
```

```
            TagOperation operation = TagOperation::None)
      : StringTagType(value, operation)
      { }
  };
```

The only change needed for LogLevel is to change the parent class from TagType to the more specific StringTagType. We no longer need to worry about specifying std::string as a template parameter because that information is built into the StringTagType class. I had wanted to keep the derived tag classes completely unchanged, but this slight modification is not so bad because there is no need to write any comparison code.

There is more work to be done in the TagType class. In the protected section at the end of the TagType class, make these changes:

```
  protected:
      TagType (ValueT const & value,
          TagOperation operation)
      : Tag(T::key, value), mValue(value), mOperation(operation)
      { }

      virtual bool compareTagTypes (ValueT const & value,
          TagOperation operation,
          ValueT const & criteria) const
      {
          return false;
      }

      ValueT mValue;
      TagOperation mOperation;
  };
```

The protected constructor needs to store the operation, and this is where the virtual compareTagTypes method is declared and given a default implementation that returns false. The TagType class also implements the match method that was declared in the Tag class, like this:

```
      bool match (Tag const & other) const override
      {
          if (key() != other.key())
          {
              return false;
```

```
        }
        TagType const & otherCast =
                static_cast<TagType const &>(other);
        if (mOperation == TagOperation::None)
        {
            switch (otherCast.mOperation)
            {
            case TagOperation::None:
                return mValue == otherCast.mValue;

            default:
                return compareTagTypes(mValue,
                    otherCast.mOperation,
                    otherCast.mValue);
            }
        }
        switch (otherCast.mOperation)
        {
        case TagOperation::None:
            return compareTagTypes(otherCast.mValue,
                mOperation,
                mValue);

        default:
            return false;
        }
    }
```

The match method first checks the keys to see whether the two tags being compared have the same key. If the keys match, then the types are assumed to be the same and the other tag is cast to the same TagType.

We have a couple of scenarios to figure out. At least one of the tags should be a normal tag without an operation and is the tag we'll call the value. The other tag can also be a regular tag without an operation, in which case all we need to do is compare the two values for equality.

If one of the two tags is normal and the other has a comparison operation other than None, then the tag with the comparison operator set is treated as the criteria. Remember that it's important to know which is the value and which is the criteria. The code needs to handle the case where we are comparing a value with a criterion or the case where we are comparing a criterion with a value. We call into the virtual `compareTagTypes` method to do the actual comparison, making sure to pass `mValue` and `otherCast.mValue` according to which is the normal tag and which is the criteria.

And finally, if both tags have the comparison operator set to something other than `None`, then we treat the match as `false` because it doesn't make sense to compare two criteria tags against each other.

There's a bit of complexity wrapped up in the `match` method that I wanted to implement in only one spot. This is why I decided to keep the `TagType` class and create value type-specific intermediate classes such as `StringTagType`. The `TagType` class implements part of the comparison by figuring out what is being compared with what and then relies on the type-specific classes to do the actual comparison.

We need to add other type-specific intermediate tag classes. All of these go in `Log.h`, right after the `StringTagType` class. Here is the one for the `int` type:

```
template <typename T>
class IntTagType : public TagType<T, int>
{
protected:
    IntTagType (int const & value,
        TagOperation operation)
    : TagType<T, int>(value, operation)
    { }

    bool compareTagTypes (int const & value,
        TagOperation operation,
        int const & criteria) const override
    {
        switch (operation)
        {
        case TagOperation::Equal:
            return value == criteria;

        case TagOperation::LessThan:
            return value < criteria;

        case TagOperation::LessThanOrEqual:
            return value <= criteria;
```

```
            case TagOperation::GreaterThan:
                return value > criteria;

            case TagOperation::GreaterThanOrEqual:
                return value >= criteria;

            default:
                return false;
        }
    }
};
```

This class is almost identical to the `StringTagType` class with changes designed for an `int` type instead of a string. Mainly, the comparisons can be done with simple arithmetic operators instead of calling the string `compare` method.

I thought about using this class for all the `int`, `long long`, and `double` arithmetic types, but that would have meant it would still need a template parameter for the actual type. Then, the question becomes one of consistency. Should the `StringTagType` class also have a template parameter to specify the type of string? Maybe. There are different kinds of strings so that almost makes sense. But what about the `bool` type? We'll need an intermediate class for Booleans too, and it seems strange to specify a `bool` template type when the class name will already have `bool` in it. So, to make everything consistent, I decided to go with separate intermediate classes for all the supported types. We'll handle ints with the `IntTagType` class and create another class called `LongLongTagType`, like this:

```
template <typename T>
class LongLongTagType : public TagType<T, long long>
{
protected:
    LongLongTagType (long long const & value,
        TagOperation operation)
    : TagType<T, long long>(value, operation)
    { }

    bool compareTagTypes (long long const & value,
        TagOperation operation,
        long long const & criteria) const override
    {
        switch (operation)
```

```
        {
        case TagOperation::Equal:
            return value == criteria;

        case TagOperation::LessThan:
            return value < criteria;

        case TagOperation::LessThanOrEqual:
            return value <= criteria;

        case TagOperation::GreaterThan:
            return value > criteria;

        case TagOperation::GreaterThanOrEqual:
            return value >= criteria;

        default:
            return false;
        }
    }
};
```

This is the class I am not very happy about because it duplicates exactly the implementation for ints. But the one thing I am happy about is the consistency it creates. It means that all the intermediate tag-type classes can be used the same way.

The next class is for doubles, and while it also has the same implementation, there is the potential to compare doubles differently because they don't compare like the integral types. There is always a little room for errors and slight discrepancies between floating-point values. For now, we're not going to do anything different about doubles, but this class will give us the ability to compare them differently if needed. The class looks like this:

```
template <typename T>
class DoubleTagType : public TagType<T, double>
{
protected:
    DoubleTagType (double const & value,
        TagOperation operation)
    : TagType<T, double>(value, operation)
```

```
    { }

    bool compareTagTypes (double const & value,
        TagOperation operation,
        double const & criteria) const override
    {
        switch (operation)
        {
        case TagOperation::Equal:
            return value == criteria;

        case TagOperation::LessThan:
            return value < criteria;

        case TagOperation::LessThanOrEqual:
            return value <= criteria;

        case TagOperation::GreaterThan:
            return value > criteria;

        case TagOperation::GreaterThanOrEqual:
            return value >= criteria;

        default:
            return false;
        }
    }
};
```

The last intermediate tag type class is for Booleans, and it does need to do something different. This class is really only interested in equality and looks like this:

```
template <typename T>
class BoolTagType : public TagType<T, bool>
{
protected:
    BoolTagType (bool const & value,
```

```
                TagOperation operation)
      : TagType<T, bool>(value, operation)
      { }


      bool compareTagTypes (bool const & value,
          TagOperation operation,
          bool const & criteria) const override
      {
          switch (operation)
          {
          case TagOperation::Equal:
              return value == criteria;

          default:
              return false;
          }
      }
  };
```

Now that we have all the tags worked out, the place where comparison needs to be made is in the `log` function, which currently uses the text of the tags to compare the normal and inverted tags. Change the `normal` block to look like this:

```
          for (auto const & normal: clause.second.normalLiterals)
          {
              // We need to make sure that the tag is
              // present and with the correct value.
              if (not activeTags.contains(normal->key()))
              {
                  allLiteralsMatch = false;
                  break;
              }
              if (not activeTags[normal->key()]->match(*normal))
              {
                  allLiteralsMatch = false;
                  break;
              }
          }
```

The code still loops through the tags and checks for the existence of the keys involved. Once it finds that the tags exist and need to be compared, instead of getting the text of each tag and comparing for equality, the code now calls the match method.

The inverted block needs to change in a similar manner, like this:

```
for (auto const & inverted:
    clause.second.invertedLiterals)
{
    // We need to make sure that the tag is either
    // not present or has a mismatched value.
    if (activeTags.contains(inverted->key()))
    {
        if (activeTags[inverted->key()]->match(
            *inverted))
        {
            allLiteralsMatch = false;
        }
        break;
    }
}
```

For the inverted loop, I was able to simplify the code a little. The real change is similar to the normal loop where the match method is called to make the comparison instead of directly comparing the tag text.

Before we can build and try out the new test, we need to update the other derived tag types in the test application. Just like how we needed to update the LogLevel tag class to use the new intermediate tag class, we need to change all the tag classes in LogTags.h. The first is the Color class, like this:

```
class Color : public MereMemo::StringTagType<Color>
{
public:
    static constexpr char key[] = "color";

    Color (std::string const & value,
        MereMemo::TagOperation operation =
            MereMemo::TagOperation::None)
    : StringTagType(value, operation)
    { }
};
```

The `Color` class is based on a string value type, just like `LogLevel`.

The `Size` tag type also uses a string and looks like this now:

```
class Size : public MereMemo::StringTagType<Size>
{
public:
    static constexpr char key[] = "size";

    Size (std::string const & value,
        MereMemo::TagOperation operation =
            MereMemo::TagOperation::None)
    : StringTagType(value, operation)
    { }
};
```

The `Count` and `Identity` tag types are based on an `int` type and a `long long` type respectively, and they look like this:

```
class Count : public MereMemo::IntTagType<Count>
{
public:
    static constexpr char key[] = "count";

    Count (int value,
        MereMemo::TagOperation operation =
            MereMemo::TagOperation::None)
    : IntTagType(value, operation)
    { }
};

class Identity : public MereMemo::LongLongTagType<Identity>
{
public:
    static constexpr char key[] = "id";

    Identity (long long value,
        MereMemo::TagOperation operation =
```

```
            MereMemo::TagOperation::None)
        : LongLongTagType(value, operation)
        { }
};
```

And finally, the `Scale` and `CacheHit` tag types are based on a `double` type and a `bool` type, and look like this:

```
class Scale : public MereMemo::DoubleTagType<Scale>
{
public:
    static constexpr char key[] = "scale";

    Scale (double value,
        MereMemo::TagOperation operation =
            MereMemo::TagOperation::None)
        : DoubleTagType(value, operation)
        { }
};

class CacheHit : public MereMemo::BoolTagType<CacheHit>
{
public:
    static constexpr char key[] = "cache_hit";

    CacheHit (bool value,
        MereMemo::TagOperation operation =
            MereMemo::TagOperation::None)
        : BoolTagType(value, operation)
        { }
};
```

The changes to each tag type were minimal. I think this is acceptable, especially because the tests that use the tag types don't need to change. Let's take another look at the test that started this section:

```
TEST("Tag values can be used to filter messages")
{
    MereTDD::SetupAndTeardown<TempFilterClause> filter;
```

```cpp
    MereMemo::addFilterLiteral(filter.id(),
        Count(100, MereMemo::TagOperation::GreaterThan));

    std::string message = "values ";
    message += Util::randomString();
    MereMemo::log(Count(1)) << message;

    bool result = Util::isTextInFile(message,
        "application.log");
    CONFIRM_FALSE(result);

    MereMemo::log() << Count(101) << message;

    result = Util::isTextInFile(message, "application.log");
    CONFIRM_FALSE(result);

    MereMemo::log(Count(101)) << message;

    result = Util::isTextInFile(message, "application.log");
    CONFIRM_TRUE(result);
}
```

This test should make more sense now. It creates a `Count` tag with a value of `100` and a `TagOperation` tag of `GreaterThan`. The operation is what makes this tag into a criteria tag that can be compared with other instances of the `Count` tag to see if the count in the other instance is really greater than 100 or not.

Then, the test tries to log with a normal `Count` tag with a value of `1`. We know now how this will fail the match, and the log message will be ignored.

The test then tries to log with a `Count` tag of `101`, but this time, the tag is outside of the `log` function and will not be considered. The second log message will also be ignored without ever trying to call `match`.

The test then tries to log with a count of `101` inside the `log` function. This one should match because 101 is indeed greater than 100, and the message should appear in the output log file.

Notice how the test is structured. It starts out with a couple of known scenarios that should not succeed before finally moving on to a scenario that should succeed. This is a good pattern for you to follow when writing your tests and helps to confirm that everything is working as designed.

The filtering is now working completely even with relative comparisons! The rest of this chapter will provide insights and advice to help you design better tests.

When is testing too much?

I remember a story I heard once about a child that was in a hospital in intensive care and was connected to all the monitoring machines, including one that watched the heartbeat electrical signals. The child's condition took a sudden turn for the worse and showed all the signs of a lack of blood flow to the brain. The doctors couldn't figure out why because the heart was beating, and they were about to send the child for a scan to look for a blood clot that would cause a stroke when one doctor thought to listen for a heartbeat. There was none. The machine showed that the heart was beating but there was no sound to confirm the beat. The doctors were able to determine that swelling around the heart was putting pressure on the heart and preventing it from beating. I don't know how, but they reduced the swelling and the child's heart started pumping again.

Why does this story come to mind? Because the machine that monitored heart activity was looking for electrical signals. In normal circumstances, the presence of proper electrical signals is a great way to monitor heart activity. But it's indirect. Electrical signals are *how* the heart beats. The signals cause the heart to beat, but as the story shows, they don't always mean *that* the heart is beating.

It's easy to fall into the same trap with software testing. We think that because we have a lot of tests, the software must be well-tested. But are the tests really testing the right things? In other words, is each test looking for tangible results? Or are some tests instead looking at how the results would normally be obtained?

When is testing too much? My answer is that testing is good, and every test that you can add will normally help to improve the quality of the software. Testing can become too much if it starts looking at the wrong things.

It's not that a test that is looking at the wrong thing is bad. The bad part comes when we rely on that test to predict some outcome. It's much better to directly confirm the desired outcome than it is to confirm some internal step along the way.

For example, have a look at a recent test that added a `filter` literal:

```
TEST("Tag values can be used to filter messages")
{
    MereTDD::SetupAndTeardown<TempFilterClause> filter;
    MereMemo::addFilterLiteral(filter.id(),
        Count(100, MereMemo::TagOperation::GreaterThan));
```

We could have verified that a filter was indeed added to the collection. We have access to call the `getFilterClauses` function from within the test and examine each clause and look for the literal just added. We could even confirm that the literal itself behaves as expected and has the value `100` assigned to the literal.

The test doesn't do this. Why? Because that is how filters work. Looking for a filter in the collection would be like watching heartbeat electrical signals. The ability to call `getFilterClauses` is a detail that exists just because of our desire to keep the logging library contained in a single header file. The function is not intended to be called by customers. The test instead looks at the results of setting the filter.

Once the filter is set, the test tries to log a few messages and makes sure that the results match the expectations.

What if the logging library needed some type of custom collection? Would it make sense to test that the filter literal was properly added to the collection then? Again, my answer is no, at least not here in the filter test.

If the project needed a custom collection, then it would need tests to make sure that the collection works. I'm not saying to skip the tests for any code that needs to be written just because that code serves a supporting role within a project. What I am saying is to keep the tests focused on what they are testing. What is the desired outcome that the test is looking to confirm? In the case of the filters test, the desired outcome is that some log messages will be ignored while others will appear in the output log file. The test directly sets up the conditions needed to confirm the outcome, causes the needed steps to be run, and confirms the outcome. Along the way, the collection and all the matching code will be tested too in an indirect manner.

If we have a custom collection involved, then indirect testing is not enough. But direct testing inside the filter test is also not appropriate. What we need is a set of tests designed to directly test the custom collection itself.

So, if we have a need for a supporting component such as a custom collection, then that component needs to be tested by itself. The tests can be included in the same overall test application. Maybe put them into their own test suite. Think about the code that will be using the component as a customer of the component and think about the customer's needs.

If the component is big enough or serves a more general purpose so that it might be useful outside of the project, then giving it a project on its own is a good idea. This is what we're doing in this book by treating the unit test library and the logging library as separate projects.

One final thought about when testing is too much will help you to identify when you are in this situation because it can be easy to slide into too much indirect testing. If you find that you need to change a lot of tests after refactoring how your software works, then you could be testing too much.

Think about how this chapter added filters and was able to keep the existing tests almost entirely unchanged. Sure—we had to change the code underneath by adding a whole set of intermediate tag-type classes, but we did not have to rewrite the existing tests.

If a refactor causes the tests to also need a lot of work, then either you are testing too much or the problem could be that you are changing the desired usage of the software. Be careful of changing how you want your design to be used, because if you are following TDD, then that initial usage is one of the first things you want to get right. Once you have the software designed in a way that makes it easy and intuitive to use, then be extra cautious about any refactoring that would cause changes to the test.

The next section explains a topic related to this section. Once you know what needs to be tested, a question that often comes up next is how to design software to make it easy to be tested, and specifically, if the tests need to reach inside the inner workings of the components being tested.

How intrusive should tests be?

There is a benefit to designing software that is easy to test. To me, this starts by following TDD and writing tests first that make use of the software as the customer would most expect it to be used. This is the most important consideration.

You don't want to make the user of your software question why extra steps are needed or why it is difficult to understand how to use your software. And by customer or user, I mean anybody that will use your software. A customer or user could be another software developer who needs to use a library that is being designed. The tests are a great example of what a user must go through. If there is an extra step that a user must take that provides no value to the user, then that step should be removed, even if the step makes it easier to test the code.

Maybe the extra step can be hidden from the user, and if so, then it might make sense to keep it as long as it makes the testing better. Anytime a test relies on something extra that the user doesn't need or know about, then the test is intruding on the software design.

I'm not saying this is a bad thing. Intrusion often has a negative meaning. It can be good for a test to be able to reach inside a component as long as you are aware that this makes it easy to fall into the trap that the previous section describes: too much testing.

The main thing to understand is that anything that a test uses should become part of a supported interface. If a component exposes an inner working so that it can be confirmed by a test, then this inner working should be accepted as part of the design and not some internal detail that is subject to change at any time.

The difference between what this section describes and the previous section comes down to what is agreed to be supported. We get into too much testing when we try to test things that either should be tested someplace else or that are internal details and should be off-limits to testing. If there's an internal detail that is stable and agreed should not change, and if that internal detail makes testing more reliable, then it might make sense for a test to use the detail.

I remember one project I worked on many years ago that exposed the internal state of classes through **Extensible Markup Language (XML)**. The state could be quite complicated at times, and using XML let the tests confirm that the state was configured correctly. The XML would then be passed to other classes that would make use of it. The user was not aware of the XML and did not need to use it, but the tests relied on it to break complicated scenarios in half. One half of the test could make sure that the configuration was correct by verifying the XML matched. The other half could then make sure that the actions taken worked properly when supplied with known XML input data.

The software did not have to be designed like this to use XML. It could even be said that the tests intruded on the design. The XML became a supported part of the design. What could have been just a detail became something more. But I would go further and say that the use of XML in this case never started out as a detail. It was a conscious design decision that was added for the specific reason of making the testing more reliable.

So far, we've only explored unit tests. That's why this book starts out by building a unit test library. When considering what should be tested and how intrusive the tests should be, there are other types of tests that the next section will begin explaining.

Where do integration or system tests go in TDD?

Sometimes, it's good to create a test that brings together multiple components and confirms that the overall system being built works as expected. These are called integration tests because they integrate multiple components to make sure they work well together. Or, the tests can be called system tests because they test the entire system. The two names are mostly interchangeable with each other.

For our microservice developer who is the target customer of the logging library, there will likely be unit tests for an individual service, and even unit tests for various classes and functions inside the service. Some of the tests for a particular service might even be called integration tests, but usually, an integration test will be working with multiple services. The services should work together to accomplish something bigger. So, having tests that make sure the overall results can be reached will help improve the reliability and quality of all the services involved.

What if you're not building a set of microservices? What if you're building a desktop application to manage a cryptocurrency wallet? You can still make use of system tests. Maybe you want a system test that opens a new wallet and makes sure it can synchronize the blockchain data up to the current block, or maybe you want another system test that stops the synchronization and then resumes it again. Each of these tests will make use of many different components such as classes and functions in the application. System tests make sure that some higher-level goal can be accomplished and, more importantly, system tests use real data that is downloaded over the network.

It's common for a system test to take a long time to complete. Add in multiple system tests, and the entire set of tests might need several hours to run. Or, maybe there are tests that continuously use the software for a day or more.

Whether or not a particular test is called a unit test or a system test often comes down to how long it takes to run and which resources are needed. Unit tests tend to be quick and are able to determine whether something passes without needing to rely on other external factors or components. If a test needs to request information from another service, then that's a good sign that the test is more of an integration test instead of a unit test. A unit test should never need to download data from a network.

When it comes to TDD, in order for a test to actually drive the design—as the name implies—then the test will normally be of the unit test variety. Don't get me wrong—system tests are important and can help uncover strange usage patterns that can be missed by unit tests. But the typical system test or integration test is not intended to make sure that the design is easy to use and intuitive. Instead, a system test makes sure that a higher-level goal can be reached and that nothing breaks the ultimate goals.

If there's any difference between system tests and integration tests, then in my mind, it comes down to integration tests being all about making sure that multiple components work well together, while a system test is more about the higher-level goals. Both integration tests and system tests are at a higher level than unit tests.

TDD makes more use of unit tests when creating the initial designs of small components and functions. And then, TDD makes use of system and integration tests to make sure that the overall solution makes sense and works properly.

You can think of all the testing we are doing for the logging library as system tests for the unit test library. We're making sure that the unit test library can actually help design another project.

As for where to put system or integration tests, they normally belong in a different test project— something that can be run on its own. This could even be a script. If you put them in the same test project as the unit tests, then there needs to be some way to only run the unit tests when a quick response is needed.

Other than system and integration tests, there are still more tests you'll want to consider adding. The next section describes more types of tests.

What about other types of tests?

There are still more types of tests to consider, such as performance testing, load testing, and penetration testing. You can even get into usability testing, upgrade testing, certification testing, continuous operation testing, and more, including types that I've probably never heard of.

Each type of test has a purpose that is valuable to software development. Each type has its own process and steps, ways of running the test, and ways to verify success.

A performance test might pick a specific scenario such as loading a large file and making sure that the operation can complete within a certain amount of time. If the test also checks to make sure that the operation completes by only using a certain amount of computer memory or CPU time, then it starts becoming more of a load test, in my opinion. And if the test makes sure that the end user doesn't have to wait or is notified of a delay, then it starts becoming more of a usability test.

The lines between the test types sometimes are not clear. The previous section already explained that system tests and integration tests are often the same thing, with a subtle distinction that often doesn't matter. The same is true of other tests. For example, whether a particular test is a load test or a performance test often comes down to the intent. Is the test trying to make sure that an operation completes in a certain time? Who decides what time is good enough? Or, is the test trying to make sure that an operation can complete while other things are going on at the same time? Or, maybe for a test that loads a large file, a large file of several megabytes is used for performance testing because that is a typical large file that a customer might encounter, while a load test would try to load a file much larger. These are just some ideas.

Penetration tests are a little different because they are normally created as part of an official security review. The whole software solution will be analyzed, lots of documents produced, and tests created. A penetration test is often trying to make sure that the software does not crash when malicious data is provided or when the system is misused.

Other penetration tests will check for information disclosure. Is it possible to misuse the software so that an attacker gains knowledge that should have remained confidential?

Even more important are penetration tests that catch data manipulation. A common example is students trying to change their grades, but this type of attack can be used to steal money or delete critical information.

Elevation-of-privilege attacks are super important to prevent penetration testing because they let an attacker gain access that can lead to more attacks. When an attacker is able to take control of a remote server, this is an obvious elevation of privilege, but elevation of privilege can be used to gain any extra permissions or capabilities that an attacker would not normally have.

Usability tests are more subjective and often involve customer interviews or trials.

All of the various different types of tests are important, and my goal with this section is not to list or describe every type of test possible but to give you an idea of the types of testing available and which benefits different tests can provide.

Software testing is not a question of which tests to use but where each type fits into the process. An entire book could be written about each of these test types, and many have been written. There's a reason this book is so focused on unit testing: because unit tests are closest to the TDD process.

Summary

The TDD process is much more important than the features added to the logging library in this chapter. We added log levels, tags, and filtering, and even refactored the design of the logging library. And while all of this is valuable, the most important thing to pay attention to is the process involved.

The reason this chapter is so detailed is so that you can see all the decisions that went into the design and how tests were used to guide the entire process. You can apply this learning to your own projects. And if you also use the logging library, then that's a bonus.

You learned the importance of understanding the needs of the customer. A customer doesn't have to be a person who walks into a store to buy something. A customer is the intended user of whatever software is being developed. This could even be another software developer or another team within your company. Understanding the needs of the intended user will let you write better tests that solve those needs.

It's very easy to write a function or design an interface that seems appropriate, only to find it difficult to use later. Writing the tests first helps to avoid usage problems. And you saw in this chapter a place where I still had to go back and change how a function worked because a test showed it to be backward.

There was an extensive set of changes needed to support filtering log messages by value, and this chapter showed how to make changes while keeping the tests unchanged.

One of the best ways to understand TDD is to use the process in a project. This chapter developed a lot of new code for the logging library to give you a front-row view into the process and gives you more than simple examples could ever show.

The next chapter will explore dependencies and will extend the logging library to send log messages to more than a single log file destination.

11

Managing Dependencies

Identifying dependencies and implementing your code around common interfaces that the dependencies use will help you in many ways. You'll be able to do the following things:

- Avoid waiting for another team or even yourself to finish a complicated and necessary component

- Isolate your code and make sure it works, even if there are bugs in other code that you use

- Achieve greater flexibility with your designs so that you can change the behavior by simply changing dependent components

- Create interfaces that clearly document and highlight essential requirements

In this chapter, you'll learn what dependencies are and how to design your code to use them. By the end of this chapter, you'll learn how to finish writing your code faster and prove that it works, even if the rest of the project is not ready for it yet.

You don't need to be using TDD to design and use dependencies. But if you are using TDD, then the whole process becomes even better because you'll also be able to write better tests that can focus on specific areas of code without worrying about extra complexity and bugs coming from outside the code.

This chapter will cover the following main topics:

- Designing with dependencies

- Adding multiple logging outputs

Technical requirements

All code in this chapter uses standard C++ that builds on any modern C++ 20 or later compiler and standard library. The code uses the testing library from *Part 1*, *Testing MVP*, of this book and continues the development of a logging library started in the previous chapters.

You can find all the code for this chapter in the following GitHub repository:

```
https://github.com/PacktPublishing/Test-Driven-Development-with-CPP
```

Designing with dependencies

Dependencies are not always obvious. If a project uses a library, such as how the logging project uses the unit test library, then that's an easy dependency to spot. The logging project depends on the unit test library to function correctly. Or in this case, only the logging tests depend on the unit test library. But that's enough to form a dependency.

Another easy dependency to spot is if you need to call another service. Even if the code checks to see if the other service is available first before making a call, the dependency still exists.

Libraries and services are good examples of *external dependencies*. You have to do extra work to get a project to use the code or services of another project, which is why an external dependency is so easy to spot.

Other dependencies are harder to spot, and these are usually *internal dependencies* within the project. In a way, almost all the code in a project depends on the rest of the code doing what it's supposed to do. So let's refine what we mean by a dependency. Normally, when a dependency is mentioned, as it relates to code design, we refer to something that can be exchanged.

This might be easiest to understand with the external service dependency example. The service operates on its own with a well-defined interface. You make a request to a service based on its location or address using the interface that the service defines. You could instead call a different service for the same request if the first service is not available. Ideally, the two services would use the same interface so that the only thing your code needs to change is the address.

If the two services use different interfaces, then it might make sense to create a wrapper for each service that knows how to translate between what each service expects and a *common interface* that your code will use. With a common interface, you can swap one service for another without changing your code. Your code depends on the service interface definition more than any specific service.

If we look at internal design decisions, maybe there is a base class and a derived class. The derived class definitely depends on the base class, but this is not the type of dependency that can be changed without rewriting the code to use a different base class.

We get closer to a dependency that can be swapped when considering the tags that the logging library defines. New tags can be defined and used without changing existing code. And the logging library can use any tag without worrying about what each tag does. But are we really swapping out tags? To me, the tags were designed to solve the problem of logging key=value elements in the log file in a consistent manner that does not depend on the data type of the value. Even though the logging library depends on tags and the interface they use, I wouldn't classify the tag design as the same type of dependency as the external service.

I mentioned early on when thinking about the logging library that we will need the ability to send the log information to different destinations, or maybe even multiple destinations. The code uses the `log` function and expects it to either be ignored or to go somewhere. The ability to send a log message to a specific destination is a dependency that the logging library needs to rely on. The logging library should let the project doing the logging decide on the destination.

And this brings us to another aspect of dependencies. A dependency is often something that is configured. What I mean is that we can say that the logging library depends on some component to perform the task of sending a message to a destination. The logging library can be designed to choose its own destination, or the logging library can be told what dependency to use. When we let other code control the dependencies, we get something called *dependency injection*. You get a more flexible solution when you let the calling code inject dependencies.

Here's some initial code that I put into the `main` function to configure a component that knows how to send log messages to a file and then inject the file component into the logger so that the logger will know where to send the log messages:

```
int main ()
{
    MereMemo::FileOutput appFile("application.log");
    appFile.maxSize() = 10'000'000;
    appFile.rolloverCount() = 5;
    MereMemo::addLogOutput(appFile);

    MereMemo::addDefaultTag(info);
    MereMemo::addDefaultTag(green);

    return MereTDD::runTests(std::cout);
}
```

The idea is to create a class called `FileOutput` and give it the name of the file to write log messages. Because we don't want log files to get too big, we should be able to specify a maximum size. The code uses 10 million bytes for the maximum size. What do we do when a log file reaches the maximum size? We should stop writing to that file and create a new file. We should be able to specify how many log files to create before we start deleting old log files. The code sets the maximum number of log files to five.

Once the `FileOutput` instance is created and configured the way we want, it is injected into the logging library by calling the `addLogOutput` function.

Will this code meet our needs? Is it intuitive and easy to understand? Even though this is not a test, we're still following TDD by concentrating on the usage of a new feature before writing the code to implement the new feature.

As for meeting our needs, that's not really the right question to ask. We need to ask if it will meet the needs of our target customer. We're designing the logging library to be used by a micro-services developer. There might be hundreds of services running on a server computer and we really should place the log files in a specific location. The first change we'll need is to let the caller specify a path where the log files should be created. The path seems like it should be separate from the filename.

And for the filenames, how will we name multiple log files? They can't all be called `application.`
`log`. Should the files be numbered? They will all be placed in the same directory and the only
requirement that the filesystem needs is that each file has a unique name. We need to let the caller
provide a pattern for the log filenames instead of a single filename. A pattern will let the logging library
know how to make the name unique while still following the overall naming style that the developer
wants. We can change the initial code to look like this instead:

```
MereMemo::FileOutput appFile("logs");
appFile.namePattern() = "application-{}.log";
appFile.maxSize() = 10'000'000;
appFile.rolloverCount() = 5;
MereMemo::addLogOutput(appFile);
```

When designing a class, it's a good idea to make the class work with reasonable defaults after
construction. For file output, the bare minimum we need is the directory to create the log files. The
other properties are nice but not required. If the name pattern is not provided, we can default to a
simple unique number. The max size can have an unlimited default, or at least a really big number. And
we only need a single log file. So, the rollover count can be some value that tells us to use a single file.

I decided to use simple open and close curly braces { } for the placeholder in the pattern where a
unique number will be placed. We'll just pick a random three-digit number to make the log filename
unique. That will give us up to a thousand log files, which should be more than enough. Most users
will only want to keep a handful and delete older files.

Because the output is a dependency that can be swapped or even have multiple outputs at the same
time, what would a different type of output look like? We'll figure out what the output dependency
component interface will be later. For now, we just want to explore how to use different outputs. Here
is how output can be sent to the `std::cout` console:

```
MereMemo::StreamOutput consoleStream(std::cout);
MereMemo::addLogOutput(consoleStream);
```

The console output is an ostream so we should be able to create a stream output that can work with
any ostream. This example creates an output component called `consoleStream`, which can be
added to the log output just like the file output.

When using TDD, it's important to avoid interesting features that may not really be needed by the
customer. We're not going to add the ability to remove outputs. Once an output is added to the logging
library, it will remain. In order to remove an output, we'd have to return some sort of identifier that
can be used to later remove the same output that was added. We did add the ability to remove filter
clauses because that ability seemed likely to be needed. Removing outputs is something that seems
unlikely for most customers.

In order to design a dependency that can be swapped for another, we'll need a common interface class that all outputs implement. The class will be called `Output` and goes in `Log.h` right before the `LogStream` class, like this:

```
class Output
{
public:
    virtual ~Output () = default;
    Output (Output const & other) = delete;
    Output (Output && other) = delete;

    virtual std::unique_ptr<Output> clone () const = 0;

    virtual void sendLine (std::string const & line) = 0;

    Output & operator = (Output const & rhs) = delete;
    Output & operator = (Output && rhs) = delete;

protected:
    Output () = default;
};
```

The only methods that are part of the interface are the `clone` and the `sendLine` methods. We'll follow a similar cloning pattern as the tags, except we're not going to use templates. The `sendLine` method will be called whenever a line of text needs to be sent to the output. The other methods make sure that nobody can construct instances of `Output` directly or copy or assign one `Output` instance to another. The `Output` class is designed to be inherited from.

We'll keep track of all the outputs that have been added with the next two functions, which go right after the `Output` class like this:

```
inline std::vector<std::unique_ptr<Output>> & getOutputs ()
{
    static std::vector<std::unique_ptr<Output>> outputs;
    return outputs;
}

inline void addLogOutput (Output const & output)
{
```

```
        auto & outputs = getOutputs();
        outputs.push_back(output.clone());
}
```

The `getOutputs` function uses a static vector of unique pointers and returns the collection when requested. The `addLogOutput` function adds a clone of the given output to the collection. This is all similar to how the default tags are handled.

One interesting use of dependencies that you should be aware of is their ability to swap out a real component for a fake component. We're adding two real components to manage the logging output. One will send output to a file and the other to the console. But you can also use dependencies if you want to make progress on your code and are waiting for another team to finish writing a needed component. Instead of waiting, make the component a dependency that you can swap out for a simpler version. The simpler version is not a real version, but it should be faster to write and let you continue making progress until the real version becomes available.

Some other testing libraries take this fake dependency ability a step further and let you create components with just a few lines of code that respond in various ways that you can control. This lets you isolate your code and make sure it behaves as it should because you can rely on the fake dependency to always behave as specified, and you no longer have to worry about bugs in the real dependency affecting the results of your tests. The common term for these fake components is *mocks*.

It doesn't matter if you are using a testing library that generates a mock for you with a few lines of code or if you are writing your own mock. Anytime you have a class that imitates another class, you have a mock.

Other than isolating your code from bugs, a mock can also help speed up your tests and improve collaboration with other teams. The speed is improved because the real code might need to spend time requesting or calculating a result, while the mock can return quickly without the need to do any real work. Collaboration with other teams is improved because everybody can agree to simple mocks that are quick to develop and can be used to communicate design changes.

The next section will implement the file and stream output classes based on the common interface. We'll be able to simplify the `LogStream` class and the `log` function to use the common interface, which will document and make it easier to understand what is really needed to send log messages to an output.

Adding multiple logging outputs

A good way to validate that a design works with multiple scenarios is to implement solutions for each scenario. We have a common `Output` interface class that defines two methods, `clone` and `sendLine`, and we need to make sure this interface will work for sending log messages to a log file and to the console.

Let's start with a class called `FileOutput` that inherits from `Output`. The new class goes in `Log.h` right after the `getOutputs` and the `addLogOutput` functions, like this:

```
class FileOutput : public Output
{
public:
    FileOutput (std::string_view dir)
    : mOutputDir(dir),
    mFileNamePattern("{}"),
    mMaxSize(0),
    mRolloverCount(0)
    { }

    FileOutput (FileOutput const & rhs)
    : mOutputDir(rhs.mOutputDir),
    mFileNamePattern(rhs.mFileNamePattern),
    mMaxSize(rhs.mMaxSize),
    mRolloverCount(rhs.mRolloverCount)
    { }

    FileOutput (FileOutput && rhs)
    : mOutputDir(rhs.mOutputDir),
    mFileNamePattern(rhs.mFileNamePattern),
    mMaxSize(rhs.mMaxSize),
    mRolloverCount(rhs.mRolloverCount),
    mFile(std::move(rhs.mFile))
    { }

    ~FileOutput ()
    {
        mFile.close();
    }

    std::unique_ptr<Output> clone () const override
    {
        return std::unique_ptr<Output>(
            new FileOutput(*this));
```

```
    }

    void sendLine (std::string const & line) override
    {
        if (not mFile.is_open())
        {
            mFile.open("application.log", std::ios::app);
        }
        mFile << line << std::endl;
        mFile.flush();
    }

protected:
    std::filesystem::path mOutputDir;
    std::string mFileNamePattern;
    std::size_t mMaxSize;
    unsigned int mRolloverCount;
    std::fstream mFile;
};
```

The `FileOutput` class follows the usage that was determined in the previous section, which looks like this:

```
MereMemo::FileOutput appFile("logs");
appFile.namePattern() = "application-{}.log";
appFile.maxSize() = 10'000'000;
appFile.rolloverCount() = 5;
MereMemo::addLogOutput(appFile);
```

We give the `FileOutput` class a directory in the constructor where the log files will be saved. The class also supports a name pattern, a max log file size, and a rollover count. All the data members need to be initialized in the constructors and we have three constructors.

The first constructor is a normal constructor that accepts the directory and gives default values to the other data members.

The second constructor is the copy constructor, and it initializes the data members based on the values in the other instance of `FileOutput`. Only the `mFile` data member is left in a default state because we don't copy fstreams.

The third constructor is the move copy constructor, and it looks almost identical to the copy constructor. The only difference is that we now move the fstream into the FileOutput class being constructed.

The destructor will close the output file. This is actually a big improvement over what was done up to this point. We used to open and close the output file each time a log message was made. We'll now open the log file and keep it open until we need to close it at a later time. The destructor makes sure that the log file gets closed if it hasn't already been closed.

Next is the clone method, which calls the copy constructor to create a new instance that gets sent back as a unique pointer to the base class.

The sendLine method is the last method, and it needs to check whether the output file has been opened already or not before sending the line to the file. We'll add the ending newline here after each line gets sent to the output file. We also flush the log file after every line, which helps to make sure that the log file contains everything written to it in case the application doing the logging crashes suddenly.

The last thing we need to do in the FileOutput class is to define the data members. We're not going to fully implement all the data members though. For example, you can see that we're still opening a file called application.log instead of following the naming pattern. We have the basic idea already and skipping the data members will let us test this part to make sure we haven't broken anything. We'll need to comment out the configuration in the main function, so it looks like this for now:

```
MereMemo::FileOutput appFile("logs");
//appFile.namePattern() = "application-{}.log";
//appFile.maxSize() = 10'000'000;
//appFile.rolloverCount() = 5;
MereMemo::addLogOutput(appFile);
```

We can always come back to the configuration methods and the directory later once we get the multiple outputs working in a basic manner. This follows the TDD practice of doing as little as possible each step along the way. In a way, what we're doing is creating a mock for the ultimate FileOutput class.

I almost forgot to mention that because we're using filesystem features, such as path, we need to include filesystem at the top of Log.h, like this:

```
#include <algorithm>
#include <chrono>
#include <ctime>
#include <filesystem>
#include <fstream>
#include <iomanip>
#include <map>
#include <memory>
```

```
#include <ostream>
#include <sstream>
#include <string>
#include <string_view>
#include <vector>
```

We'll make use of filesystem more once we start rolling log files over into new files instead of always opening the same file each time.

Next is the StreamOutput class, which can go in Log.h right after the FileOutput class and looks like this:

```
class StreamOutput : public Output
{
public:
    StreamOutput (std::ostream & stream)
    : mStream(stream)
    { }

    StreamOutput (StreamOutput const & rhs)
    : mStream(rhs.mStream)
    { }

    std::unique_ptr<Output> clone () const override
    {
        return std::unique_ptr<Output>(
            new StreamOutput(*this));
    }

    void sendLine (std::string const & line) override
    {
        mStream << line << std::endl;
    }

protected:
    std::ostream & mStream;
};
```

The `StreamOutput` class is simpler than the `FileOutput` class because it has fewer data members. We only need to keep track of an ostream reference that gets passed in the constructor in `main`. We also don't need to worry about a specific move copy constructor because we can easily copy the ostream reference. The `StreamOutput` class was already added in `main` like this:

```
MereMemo::StreamOutput consoleStream(std::cout);
MereMemo::addLogOutput(consoleStream);
```

The `StreamOutput` class will hold a reference to `std::cout` that `main` passes to it.

Now that we're working with the output interface, we no longer need to manage a file in the `LogStream` class. The constructors can be simplified to no longer worry about an fstream data member, like this:

```
LogStream ()
: mProceed(true)
{ }

LogStream (LogStream const & other) = delete;

LogStream (LogStream && other)
: std::stringstream(std::move(other)),
mProceed(other.mProceed)
{ }
```

The destructor of the `LogStream` class is where all the work happens. It no longer needs to send the message directly to a file that the class manages. The destructor now gets *all* the outputs and sends the message to each one using the common interface, like this:

```
~LogStream ()
{
    if (not mProceed)
    {
        return;
    }

    auto & outputs = getOutputs();
    for (auto const & output: outputs)
    {
        output->sendLine(this->str());
    }
}
```

Remember that the LogStream class inherits from std::stringstream and holds the message to be logged. If we are to proceed, then we can get the fully formatted message by calling the str method.

The end of LogStream no longer needs the mFile data member and only needs the mProceed flag, like this:

```
private:
    bool mProceed;
};
```

Because we removed the LogStream constructor parameters for the filename and open mode, we can simplify how the LogStream class gets created in the log function like this:

```
inline LogStream log (std::vector<Tag const *> tags = {})
{
    auto const now = std::chrono::system_clock::now();
    std::time_t const tmNow =
        std::chrono::system_clock::to_time_t(now);
    auto const ms = duration_cast<std::chrono::milliseconds>(
        now.time_since_epoch()) % 1000;

    LogStream ls;
    ls << std::put_time(std::gmtime(&tmNow),
        "%Y-%m-%dT%H:%M:%S.")
        << std::setw(3) << std::setfill('0')
        << std::to_string(ms.count());
```

We can now construct the ls instance without any arguments, and it will use all the outputs that have been added.

Let's check the test application by building and running the project. The output to the console looks like this:

```
Running 1 test suites
-------------- Suite: Single Tests
------- Test: Message can be tagged in log
2022-07-24T22:32:13.116 color="green" log_level="error" simple
7809
Passed
------- Test: log needs no namespace when used with LogLevel
2022-07-24T22:32:13.118 color="green" log_level="error" no
```

```
namespace
Passed
------- Test: Default tags set in main appear in log
2022-07-24T22:32:13.118 color="green" log_level="info" default
tag 9055
Passed
------- Test: Multiple tags can be used in log
2022-07-24T22:32:13.118 color="red" log_level="debug"
size="large" multi tags 7933
Passed
------- Test: Tags can be streamed to log
2022-07-24T22:32:13.118 color="green" log_level="info" count=1
1 type 3247
2022-07-24T22:32:13.118 color="green" log_level="info"
id=123456789012345 2 type 6480
2022-07-24T22:32:13.118 color="green" log_level="info"
scale=1.500000 3 type 6881
2022-07-24T22:32:13.119 color="green" log_level="info" cache_
hit=false 4 type 778
Passed
------- Test: Tags can be used to filter messages
2022-07-24T22:32:13.119 color="green" log_level="info" filter
1521
Passed
------- Test: Overridden default tag not used to filter
messages
Passed
------- Test: Inverted tag can be used to filter messages
Passed
------- Test: Tag values can be used to filter messages
2022-07-24T22:32:13.119 color="green" count=101 log_
level="info" values 8461
Passed
------- Test: Simple message can be logged
2022-07-24T22:32:13.120 color="green" log_level="info" simple
9466 with more text.
Passed
------- Test: Complicated message can be logged
```

```
2022-07-24T22:32:13.120 color="green" log_level="info"
complicated 9198 double=3.14 quoted="in quotes"
Passed
----------------------------------
Tests passed: 11
Tests failed: 0
```

You can see that the log messages did indeed go to the console window. The log messages are included in the console alongside the test results. What about the log file? It looks like this:

```
2022-07-24T22:32:13.116 color="green" log_level="error" simple
7809
2022-07-24T22:32:13.118 color="green" log_level="error" no
namespace
2022-07-24T22:32:13.118 color="green" log_level="info" default
tag 9055
2022-07-24T22:32:13.118 color="red" log_level="debug"
size="large" multi tags 7933
2022-07-24T22:32:13.118 color="green" log_level="info" count=1
1 type 3247
2022-07-24T22:32:13.118 color="green" log_level="info"
id=123456789012345 2 type 6480
2022-07-24T22:32:13.118 color="green" log_level="info"
scale=1.500000 3 type 6881
2022-07-24T22:32:13.119 color="green" log_level="info" cache_
hit=false 4 type 778
2022-07-24T22:32:13.119 color="green" log_level="info" filter
1521
2022-07-24T22:32:13.119 color="green" count=101 log_
level="info" values 8461
2022-07-24T22:32:13.120 color="green" log_level="info" simple
9466 with more text.
2022-07-24T22:32:13.120 color="green" log_level="info"
complicated 9198 double=3.14 quoted="in quotes"
```

The log file contains only the log messages, which are the same log messages that were sent to the console window. This shows that we have multiple outputs! There's no good way to verify that the log messages are being sent to the console window, such as how we can open the log file and search for a specific line.

But we could add yet another output using the `StreamOutput` class that is given `std::fstream` instead of `std::cout`. We can do this because fstream implements ostream, which is all that the `StreamOutput` class needs. This is also dependency injection because the `StreamOutput` class depends on an ostream and we can give it any ostream we want it to use, like this:

```
#include <fstream>
#include <iostream>

int main ()
{
    MereMemo::FileOutput appFile("logs");
    //appFile.namePattern() = "application-{}.log";
    //appFile.maxSize() = 10'000'000;
    //appFile.rolloverCount() = 5;
    MereMemo::addLogOutput(appFile);

    MereMemo::StreamOutput consoleStream(std::cout);
    MereMemo::addLogOutput(consoleStream);

    std::fstream streamedFile("stream.log", std::ios::app);
    MereMemo::StreamOutput fileStream(streamedFile);
    MereMemo::addLogOutput(fileStream);

    MereMemo::addDefaultTag(info);
    MereMemo::addDefaultTag(green);

    return MereTDD::runTests(std::cout);
}
```

We're not going to make this change. It's just for demonstration purposes only. But it shows that you can open a file and pass that file to the `StreamOutput` class to use instead of the console output. If you do make this change, then you'll see that the `stream.log` and `application.log` files are the same.

Why would you want to consider using `StreamOutput` as if it was `FileOutput`? And why do we need `FileOutput` if `StreamOutput` can also write to a file?

First of all, `FileOutput` is specialized for files. It will eventually know how to check the current file size to make sure it doesn't get too big and roll over to a new log file whenever the current log file approaches the maximum size. There is a need for file management that `StreamOutput` will not even be aware of.

The `StreamOutput` class is simpler though because it doesn't need to worry about files at all. You might want to use `StreamOutput` to write to a file in case the `FileOutput` class takes too long to create. Sure, we created a simplified `FileOutput` without all the file management features, but another team might not be so willing to give you a partial implementation. You might find it better to use a mock solution while you wait for a full implementation.

The ability to swap one implementation for another is a big advantage you get with properly managed dependencies.

In fact, this book will leave the current implementation of `FileOutput` as it is now because finishing the implementation would take us into topics that have little to do with learning about TDD.

Summary

We not only added a great new feature to the logging library that lets it send log messages to multiple destinations but we also added this ability using an interface. The interface helps document and isolate the idea of sending lines of text to a destination. This helped uncover a dependency that the logging library has. The logging library depends on the ability to send text somewhere.

The destination could be a log file or the console, or somewhere else. Until we identified this dependency, the logging library was making assumptions in many places that it was working with a log file only. We were able to simplify the design and, at the same time create a more flexible design.

We were also able to get the file logging working without a complete file logging component. We created a mock of the file logging component that leaves out all the additional file management tasks that a full implementation will need. While useful, the additional capabilities are not needed right now, and the mock will let us proceed without them.

The next chapter will go back to the unit testing library and will show you how to enhance the confirmations to a new style that is extensible and easier to understand.

Part 3:
Extending the TDD Library to Support the Growing Needs of the Logging Library

This book is divided into three parts. In this third and final part, we'll be enhancing the unit test confirmations to use a new modern style called Hamcrest confirmations. You'll also learn how to test services and how to test with multiple threads. This third part will tie everything you've learned so far together and prepare you to use TDD in your own projects.

The following chapters are covered in this part:

- *Chapter 12, Creating better test assertions*
- *Chapter 13, How to Test Floating-Point and Custom Values*
- *Chapter 14, How to Test Services*
- *Chapter 15, Testing Across Multiple Threads*

12
Creating Better Test Confirmations

This chapter introduces *Part 3*, where we extend the TDD library to support the growing needs of the logging library. *Part 1*, *Testing MVP*, of this book developed a basic unit test library, and *Part 2*, *Logging Library*, started to use the unit test library to build a logging library. Now we are following TDD, which encourages enhancing something once the basic tests are working.

Well, we managed to get a basic unit test library working and proved its worth by building a logging library. In a way, the logging library is like systems tests for the unit test library. Now it's time to enhance the unit test library.

This chapter adds a completely new type of confirmation to the unit test library. First, we'll look at the existing confirmations to understand how they can be improved and what the new solution will look like.

The new confirmations will be more intuitive, more flexible, and extensible. And remember to pay attention not only to the code being developed in this chapter but also to the process. That's because we'll be using TDD throughout to write some tests, starting with a simple solution and then enhancing the tests to create an even better solution.

In this chapter, we will cover the following main topics:

- The problem with the current confirmations
- How to simplify string confirmations
- Enhancing the unit test library to support Hamcrest-style confirmations
- Adding more Hamcrest matcher types

Technical requirements

All code in this chapter uses standard C++ that builds on any modern C++ 20 or later compiler and standard library. The code is based on and continues enhancing the testing library from *Part 1, Testing MVP*, of this book.

You can find all the code for this chapter in the following GitHub repository:

https://github.com/PacktPublishing/Test-Driven-Development-with-CPP

The problem with the current confirmations

Before we begin making changes, we should have some idea of why. TDD is all about the customer experience. How can we design something that is easy and intuitive to use? Let's start by taking a look at a couple of existing tests:

```
TEST("Test string and string literal confirms")
{
    std::string result = "abc";
    CONFIRM("abc", result);
}


TEST("Test float confirms")
{
    float f1 = 0.1f;
    float f2 = 0.2f;
    float sum = f1 + f2;
    float expected = 0.3f;
    CONFIRM(expected, sum);
}
```

These tests have served well and are easy, right? What we're looking at here is not the tests themselves but the confirmations. This style of confirmation is called the *classic style*.

How would we speak or read aloud the first confirmation? It might go like this: "*Confirm the expected value of abc matches the value of result.*"

That's not too bad, but it's a bit awkward. That's not how a person would normally talk. Without looking at any code, a more natural way to say the same thing would be: "*Confirm that result equals abc.*"

At first glance, maybe all we need to do is reverse the order of the parameters and put the actual value first followed by the expected value. But there's a piece missing. How do we know that a confirmation is checking for equality? We know because that's the only thing the existing `confirm` functions know how to check. Also, that means the `CONFIRM` macro only knows how to check for equality, too.

We have a better solution for bool values because we created special `CONFIRM_TRUE` and `CONFIRM_FALSE` macros that are easy to use and understand. And because the bool versions only take a single parameter, there's no question of expected versus actual ordering.

There's a better solution that aligns with the more natural way we would speak about a confirmation. The better solution uses something called matchers and is referred to as the *Hamcrest style*. The name "Hamcrest" is just a reordering of the letters in the word "matchers." Here is what a test would look like written in the Hamcrest style:

```
TEST("Test can use hamcrest style confirm")
{
    int ten = 10;
    CONFIRM_THAT(ten, Equals(10));
}
```

We're not really designing the Hamcrest style in this book. The style already exists and is common in other testing libraries. And the main reason the testing library in this book puts the expected value first followed by the actual value for the classical style is to follow the common practice of the classical style.

Imagine you were to reinvent a better light switch. And I've been in buildings that have tried. The light switch might actually be better in some way. But if it doesn't follow normal expectations, then people will get confused.

The same is true of the classical confirmations we started with in this book. I could have designed the confirmations to put the actual value first and maybe that would be better. But it would be unexpected for anybody who is even a little familiar with the existing testing libraries.

This brings up a great point to consider when creating designs using TDD. Sometimes, an inferior solution is better when that's what the customer expects. Remember that whatever we design should be easy and intuitive to use. The goal is not to make the ultimate and most modern design but to make something that the user will be happy with.

This is why Hamcrest matchers work. The design doesn't just switch the order of the expected and actual values because switching the order by itself would only confuse users.

Hamcrest works well because something else was added: the matcher. Notice the `Equals(10)` part of the confirmation. `Equals` is a matcher that makes it clear what the confirmation is doing. The matcher, together with a more intuitive ordering, gives a solution enough benefits to overcome the natural reluctance people have with switching to a new way of doing things. The Hamcrest style is not just a better light switch. Hamcrest is different enough and provides enough value that it avoids the confusion of a slightly better but different solution.

Also, notice that the name of the macro has changed from CONFIRM to CONFIRM_THAT. The name change is another way to avoid confusion and lets users continue to use the older classical style or opt for the newer Hamcrest style.

Now that we have a place to specify something such as Equals, we can also use different matchers such as GreaterThan or BeginsWith. Imagine that you wanted to confirm that some text begins with some expected characters. How would you write a test like that using classical confirmations? You would have to check for the beginning text outside of the confirmation and then confirm the result of the check. With the Hamcrest style and an appropriate matcher, you can confirm the text with a single-line confirmation. And you get the benefit of a more readable confirmation that makes it clear what is being confirmed.

What if you can't find a matcher that fits your needs? You can always write your own to do exactly what you need. So, Hamcrest is extensible.

Before diving into the new Hamcrest design, the next section will take a slight detour to explain an improvement to the existing classic confirm template function. This improvement will be used in the Hamcrest design, so understanding the improvement first will help later when we get to the Hamcrest code explanations.

Simplifying string confirmations

While I was writing the code for this chapter, I ran into a problem confirming string data types that reminded me of how we added support for confirming strings in *Chapter 5, Adding More Confirm Types*. The motivating factor from *Chapter 5* was to get the code to compile because we can't pass std::string to a std::to_string function. I'll briefly explain the problem again here.

I'm not sure of the exact reasons, but I think that the C++ standard library designers felt there was no need to provide an overload of std::to_string that accepts std::string because no conversion is needed. A string is already a string! Why convert something into what it already is?

Maybe this decision was on purpose or maybe it was an oversight. But it sure would have helped to have a string conversion into a string for template functions that need to convert their generic types into strings. That's because, without the overload, we have to take extra steps to avoid compile errors. What we need is a to_string function that can convert any type into a string even if the type is already a string. If we had this ability to always be able to convert types into strings, then a template wouldn't need to be specialized for strings.

In *Chapter 5*, we introduced this template:

```
template <typename T>
void confirm (
    T const & expected,
    T const & actual,
```

```
        int line)
{

    if (actual != expected)
    {
        throw ActualConfirmException(
            std::to_string(expected),
            std::to_string(actual),
            line);
    }
}
```

The `confirm` function accepts two templatized parameters, called `expected` and `actual`, that are compared for equality. If they are not equal, then the function passes both parameters to an exception that gets thrown. The parameters need to be converted into strings, as needed, by the `ActualConfirmException` constructor.

This is where we run into a problem. If the `confirm` template function is called with strings, then it doesn't compile because strings can't be converted into strings by calling `std::to_string`.

The solution we went with in *Chapter 5* was to overload the `confirm` function with a non-template version that directly accepted strings. We actually created two overloads, one for strings and one for string views. This solved the problem but left us with the following two additional overloads:

```
inline void confirm (
    std::string_view expected,
    std::string_view actual,
    int line)
{

    if (actual != expected)
    {
        throw ActualConfirmException(
            expected,
            actual,
            line);
    }
}

inline void confirm (
    std::string const & expected,
```

```
        std::string const & actual,
        int line)
{

    confirm(
        std::string_view(expected),
        std::string_view(actual),
        line);
}
```

When calling `confirm` with strings, these overloads are used instead of the template. The version that accepts `std::string` types calls into the version that takes `std::string_view` types, which uses the `expected` and `actual` parameters directly without trying to call `std::to_string`.

At the time, this wasn't such a bad solution because we already had extra overloads of `confirm` for bools and the various floating point types. Two more overloads for strings was okay. Later, you'll see how a small change will let us remove these two string overloads.

And now we come back to the problem of converting string data types with the new Hamcrest design we'll be working on in this chapter. We will no longer need extra overloads of `confirm` even for bool or floating point types. As I was working on a new solution, I came back to the earlier solution from *Chapter 5* and decided to refactor the existing classic confirms so that both solutions would be similar.

We'll get into the new design later in this chapter. But so that we don't have to interrupt that explanation, I have decided to take the detour now and explain how to remove the need for the string and string view overloads of the classic `confirm` function. Going through the explanation now should also make it easier to understand the new Hamcrest design since you'll already be familiar with this part of the solution.

Also, I'd like to add that TDD helps with this type of refactoring. Because we already have existing tests for the classic confirms, we can remove the string overloads of `confirm` and make sure that all the tests continue to pass. I've worked on projects before where only the new code would use the better solution and we would have to leave the existing code unchanged in order to avoid introducing bugs. Doing this just makes the code harder to maintain because now there would be two different solutions in the same project. Having good tests helps give you the confidence needed to change existing code.

Okay, the core of the problem is that the C++ standard library does not include overloads of `to_string` that work with strings. And while it might be tempting to just add our own version of `to_string` to the `std` namespace, this is not allowed. It would probably work, and I'm sure that lots of people have done this. But it's technically undefined behavior to add any function into the `std` namespace. There are very specific cases where we are allowed to add something into the `std` namespace, and unfortunately, this is not one of the allowed exceptions to the rules.

We will need our own version of `to_string`. We just can't put our version in the `std` namespace. That's a problem because when we call `to_string`, we currently specify the namespace by calling `std::to_string`. What we need to do is simply call `to_string` without any namespace and let the compiler look in either the `std` namespace to find the versions of `to_string` that work with numeric types, or look in our namespace to find our new version that works with strings. The new `to_string` function and the modified `confirm` template function look like this:

```
inline std::string to_string (std::string const & str)
{
    return str;
}


template <typename ExpectedT, typename ActualT>
void confirm (
    ExpectedT const & expected,
    ActualT const & actual,
    int line)
{
    using std::to_string;
    using MereTDD::to_string;
    if (actual != expected)
    {
        throw ActualConfirmException(
            to_string(expected),
            to_string(actual),
            line);
    }
}
```

We can remove the two overloads of `confirm` that take string views and strings. Now, the `confirm` template function will work for strings.

With the new `to_string` function that accepts `std::string`, all it needs to do is return the same string. We don't really need another `to_string` function that works with string views.

The `confirm` template function is a little more complicated because it now needs two types, `ExpectedT` and `ActualT`. The two types are needed for those cases when we need to compare a string literal with a string, such as in the following test:

```
TEST("Test string and string literal confirms")
{
    std::string result = "abc";
```

```
        CONFIRM("abc", result);
}
```

The reason this test used to compile when we had only a single confirm template parameter is that it wasn't calling into the template. The compiler was converting the "abc" string literal into a string and calling the overload of confirm that accepted two strings. Or maybe it was converting both the string literal and the string into string views and calling the overload of confirm that accepted two string views. Either way, because we had separate overloads of confirm, the compiler was able to make it work.

Now that we removed both confirm overloads that deal with strings, we have only the template, and we need to let it accept different types in order to compile. I know, we still have overloads for bool and the floating point types. I'm only talking about the string overloads that we can remove.

In the new template, you can see that we call to_string without any namespace specification. The compiler is able to find the versions of to_string it needs because of the two using statements inside the template function. The first using statement tells the compiler that it should consider all the to_string overloads in the std namespace. And the second using statement tells the compiler to also consider any to_string functions it finds in the MereTDD namespace.

The compiler is now able to find a version of to_string that works with the numeric types when confirm is called with numeric types. And the compiler can find our new to_string function that works with strings when needed. We no longer need to limit the compiler to only look in the std namespace.

Now we can go back to the new Hamcrest style design, which we will do in the next section. The Hamcrest design will, eventually, use a solution similar to what was just described here.

Enhancing the test library to support Hamcrest matchers

Once you get a basic implementation working and passing the tests, TDD guides us to enhance the design by creating more tests and then getting the new tests to pass. That's exactly what this chapter is all about. We're enhancing the classic style confirmations to support the Hamcrest style.

Let's start by creating a new file, called Hamcrest.cpp, in the tests folder. Now, the overall project structure should look like this:

```
MereTDD project root folder
    Test.h
    tests folder
        main.cpp
        Confirm.cpp
        Creation.cpp
```

```
Hamcrest.cpp
Setup.cpp
```

If you've been following all the code in this book so far, remember that we're going back to the *MereTDD* project that we last worked on in *Chapter 7, Test Setup and Teardown*. This is not the *MereMemo* logging project.

The Hamcrest style test that we need to support goes inside Hamcrest.cpp so that the new file looks like this:

```
#include "../Test.h"


TEST("Test can use hamcrest style confirm")
{
    int ten = 10;
    CONFIRM_THAT(ten, Equals(10));
}
```

We might as well start with the new CONFIRM_THAT macro, which goes at the end of Test.h right after the other CONFIRM macros, like this:

```
#define CONFIRM_FALSE( actual ) \
    MereTDD::confirm(false, actual, __LINE__)
#define CONFIRM_TRUE( actual ) \
    MereTDD::confirm(true, actual, __LINE__)
#define CONFIRM( expected, actual ) \
    MereTDD::confirm(expected, actual, __LINE__)
#define CONFIRM_THAT( actual, matcher ) \
    MereTDD::confirm_that(actual, matcher, __LINE__)
```

The CONFIRM_THAT macro is similar to the CONFIRM macro except that the actual parameter comes first, and instead of an expected parameter, we have a parameter called matcher. We'll also call a new function called confirm_that. The new function helps make it simpler to keep the classic style confirm overloads separate from the Hamcrest-style confirm_that function.

We don't need all the overloads in the same way that we needed for confirm. The confirm_that function can be implemented with a single template function. Place this new template in Test.h right after the classic confirm template function. Both template functions should look like this:

```
template <typename ExpectedT, typename ActualT>
void confirm (
```

```
        ExpectedT const & expected,
        ActualT const & actual,
        int line)
    {
        using std::to_string;
        using MereTDD::to_string;
        if (actual != expected)
        {
            throw ActualConfirmException(
                to_string(expected),
                to_string(actual),
                line);
        }
    }

    template <typename ActualT, typename MatcherT>
    inline void confirm_that (
        ActualT const & actual,
        MatcherT const & matcher,
        int line)
    {
        using std::to_string;
        using MereTDD::to_string;
        if (not matcher.pass(actual))
        {
            throw ActualConfirmException(
                to_string(matcher),
                to_string(actual),
                line);
        }
    }
```

We're only adding the `confirm_that` function. I decided to show both functions so that you can see the differences easier. Notice that now, the `ActualT` type is given first. The order doesn't really matter, but I like to put the template parameters in a reasonable order. We no longer have an `ExpectedT` type; instead, we have a `MatcherT` type.

The name of the new template function is different too, so there is no ambiguity due to the similar template parameters. The new template function is called confirm_that.

While the classic confirm function compares the actual parameter directly with the expected parameter, the new confirm_that function calls into a pass method on the matcher to perform the check. We don't really know what the matcher will be doing in the pass method because that is for the matcher to decide. And because any changes in the comparison from one type to another are wrapped up in the matcher, we don't need to overload the confirm_that function like we had to do for the classic confirm function. We'll still need a special code, but the differences will be handled by the matcher in this design.

This is where I realized that there needs to be a different solution for converting the matcher and actual parameters into strings. It seems pointless to override confirm_that just to avoid calling to_string when the type of ActualT is a string. So, I stopped calling std::to_string(actual) and instead started calling to_string(actual). In order for the compiler to find the necessary to_string functions, the using statements are needed. This is the explanation that the previous section describes for simplifying the string comparisons.

Now that we have the confirm_that template, we can focus on the matcher. We need to be able to call a pass method and convert a matcher into a string. Let's create a base class for all the matchers to inherit from, so they will all have a common interface. Place this base class and to_string function right after the confirm_that function in Test.h, as follows:

```
class Matcher
{
public:
    virtual ~Matcher () = default;
    Matcher (Matcher const & other) = delete;
    Matcher (Matcher && other) = delete;

    virtual std::string to_string () const = 0;

    Matcher & operator = (Matcher const & rhs) = delete;
    Matcher & operator = (Matcher && rhs) = delete;

protected:
    Matcher () = default;
};


inline std::string to_string (Matcher const & matcher)
{
```

```
        return matcher.to_string();
    }
```

The `to_string` function will let us convert a matcher into a string by calling the virtual `to_string` method in the `Matcher` base class. Notice there is no `pass` method in the `Matcher` class.

The `Matcher` class itself is a base class that doesn't need to be copied or assigned. The only common interface the `Matcher` class defines is a `to_string` method that all matchers will implement to convert themselves into a string that can be sent to the test run summary report.

What happened to the `pass` method? Well, the `pass` method needs to accept the actual type that will be used to determine whether the actual value matches the expected value. The expected value itself will be held in the derived matcher class. The actual value will be passed to the `pass` method.

The types of values accepted for the actual and expected values will be fully under the control of the derived matcher class. Because the types can change from one usage of a matcher to another, we can't define a `pass` method in the `Matcher` base class. This is okay because the `confirm_that` template doesn't work with the `Matcher` base class. The `confirm_that` template will have knowledge of the real matcher-derived class and can call the `pass` method directly as a non-virtual method.

The `to_string` method is different because we want to call the virtual `Matcher::to_string` method from within the `to_string` helper function that accepts any `Matcher` reference.

So, when converting a matcher into a string, we treat all matchers the same and go through the virtual `to_string` method. And when calling `pass`, we work directly with the real matcher class and call `pass` directly.

Let's see what a real matcher class will look like. The test we are implementing uses a matcher called `Equals`. We can create the derived `Equals` class right after the `Matcher` class and the `to_string` function, as follows:

```
template <typename T>
class Equals : public Matcher
{
public:
    Equals (T const & expected)
    : mExpected(expected)
    { }

    bool pass (T const & actual) const
    {
        return actual == mExpected;
    }
```

```
    std::string to_string () const override
    {
        using std::to_string;
        using MereTDD::to_string;
        return to_string(mExpected);
    }

private:
    T mExpected;
};
```

The `Equals` class is another template because it needs to hold the proper expected value type, and it needs to use the same type in the `pass` method for the `actual` parameter.

Notice that the `to_string` override method uses the same solution to convert the `mExpected` data member into a string that we've been using. We call `to_string` and let the compiler find an appropriate match in either the `std` or `MereTDD` namespaces.

We need one more small change to get everything working. In our Hamcrest test, we use the `Equals` matcher without any namespace specification. We could refer to it as `MereTDD::Equals`. But the namespace specification distracts from the readability of the tests. Let's add a `using namespace MereTDD` statement to the top of any test file that will use Hamcrest matchers so we can refer to them directly, like this:

```
#include "../Test.h"

using namespace MereTDD;

TEST("Test can use hamcrest style confirm")
{
    int ten = 10;
    CONFIRM_THAT(ten, Equals(10));
}
```

That's everything needed to support our first Hamcrest matcher unit test – building and running show that all tests pass. What about an expected failure? First, let's create a new test like this:

```
TEST("Test hamcrest style confirm failure")
{
```

```
    int ten = 10;
    CONFIRM_THAT(ten, Equals(9));
}
```

This test is designed to fail because 10 will not equal 9. We need to build and run once just to get the failure message from the summary report. Then, we can add a call to setExpectedFailureReason with the exactly formatted failure message. Remember that the failure message needs to match exactly, including all the spaces and punctuation. I know this can be tedious, but it should not be a test that you need to worry about unless you're testing one of your own custom matchers to make sure the custom matcher is able to format a proper error message.

After getting the exact error message, we can modify the test to turn it into an expected failure, as follows:

```
TEST("Test hamcrest style confirm failure")
{
    std::string reason = "    Expected: 9\n";
    reason += "    Actual  : 10";
    setExpectedFailureReason(reason);

    int ten = 10;
    CONFIRM_THAT(ten, Equals(9));
}
```

Building and running again shows both Hamcrest tests results, as follows:

```
------- Test: Test can use hamcrest style confirm
Passed
------- Test: Test hamcrest style confirm failure
Expected failure
    Expected: 9
    Actual  : 10
```

This is a good start. We haven't yet started talking about how to design custom matchers. Before we start custom matchers, what about other basic types? We only have a couple of Hamcrest tests that compare int values. The next section will explore other basic types and add more tests.

Adding more Hamcrest types

There is a pattern to using TDD that you should be familiar with by now. We add a little bit of something, get it working, and then add more. We have the ability to confirm int values with a Hamcrest Equals matcher. Now it's time to add more types. Some of these types might work without any extra work due to the template `confirm_that` function. Other types might need changes. We'll find out what needs to be done by writing some tests.

The first test ensures that the other integer types work as expected. Add this test to `Hamcrest.cpp`:

```
TEST("Test other hamcrest style integer confirms")
{
    char c1 = 'A';
    char c2 = 'A';
    CONFIRM_THAT(c1, Equals(c2));
    CONFIRM_THAT(c1, Equals('A'));

    short s1 = 10;
    short s2 = 10;
    CONFIRM_THAT(s1, Equals(s2));
    CONFIRM_THAT(s1, Equals(10));

    unsigned int ui1 = 3'000'000'000;
    unsigned int ui2 = 3'000'000'000;
    CONFIRM_THAT(ui1, Equals(ui2));
    CONFIRM_THAT(ui1, Equals(3'000'000'000));

    long long ll1 = 5'000'000'000'000LL;
    long long ll2 = 5'000'000'000'000LL;
    CONFIRM_THAT(ll1, Equals(ll2));
    CONFIRM_THAT(ll1, Equals(5'000'000'000'000LL));
}
```

First, the test declares a couple of chars and uses the `Equals` matcher in a couple of different ways. The first is to test for equality with another char. The second uses a char literal value, `'A'`, for the comparison.

The second set of confirmations is based on short ints. We use the `Equals` matcher with another short int and then an int literal value of `10` for the comparison.

The third set of confirmations is based on unsigned ints and, again, tries to use the `Equals` matcher with another variable of the same type and with a literal int.

The fourth set of confirmations makes sure that long long types are supported.

We're not creating helper functions designed to simulate other software being tested. You already know how to use confirmations in a real project based on the tests in the logging library. That's why this test makes things simple and just focuses on making sure that the CONFIRM_THAT macro, which calls the confirm_that template function, works.

Building and running these tests show that all tests pass with no changes or enhancements needed.

What about bool types? Here is a test that goes into Hamcrest.cpp to test bool types:

```
TEST("Test hamcrest style bool confirms")
{
    bool b1 = true;
    bool b2 = true;
    CONFIRM_THAT(b1, Equals(b2));

    // This works but probably won't be used much.
    CONFIRM_THAT(b1, Equals(true));

    // When checking a bool variable for a known value,
    // the classic style is probably better.
    CONFIRM_TRUE(b1);
}
```

This test shows that the Hamcrest style works for bool types, too. When comparing one bool variable with another, the Hamcrest style is better than the classic style. However, when comparing a bool variable with an expected true or false literal, it's actually more readable to use the classic style because we have simplified CONFIRM_TRUE and CONFIRM_FALSE macros.

Now, let's move on to strings with this test that goes into Hamcrest.cpp. Note that this test will fail to compile at first and that's okay. The test looks like this:

```
TEST("Test hamcrest style string confirms")
{
    std::string s1 = "abc";
    std::string s2 = "abc";
    CONFIRM_THAT(s1, Equals(s2));      // string vs. string
    CONFIRM_THAT(s1, Equals("abc"));   // string vs. literal
    CONFIRM_THAT("abc", Equals(s1));   // literal vs. string
}
```

There are several confirms in this test, and that's okay because they're all related. The comments help to clarify what each confirmation is testing.

We're always looking for two things with a new test. The first is whether the test compiles at all. And the second is whether it passes. Right now, the test will fail to compile with an error similar to the following:

```
MereTDD/tests/../Test.h: In instantiation of
 'MereTDD::Equals<T>::Equals(const T&) [with T = char [4]]':
MereTDD/tests/Hamcrest.cpp:63:5:    required from here
MereTDD/tests/../Test.h:209:7: error: array used as initializer
  209 |      : mExpected(expected)
      |        ^~~~~~~~~~~~~~~~~~~~
```

You might get different line numbers, so I'll explain what the error is referring to. The failure is in the Equals constructor, which looks like this:

```
    Equals (T const & expected)
    : mExpected(expected)
    { }
```

And line 63 in Hamcrest.cpp is the following line:

```
    CONFIRM_THAT(s1, Equals("abc"));   // string vs. literal
```

We're trying to construct an Equals matcher given the "abc" string literal, and this fails to compile. The reason is that the T type is an array that needs to be initialized in a different manner.

What we need is a special version of Equals that works with string literals. Since a string literal is an array of constant chars, the following template specialization will work. Place this new template in Test.h right after the existing Equals template:

```
template <typename T, std::size_t N> requires (
    std::is_same<char, std::remove_const_t<T>>::value)
class Equals<T[N]> : public Matcher
{
public:
    Equals (char const (& expected)[N])
    {
        memcpy(mExpected, expected, N);
    }

    bool pass (std::string const & actual) const
```

```
        {
            return actual == mExpected;
        }

        std::string to_string () const override
        {
            return std::string(mExpected);
        }

    private:
        char mExpected[N];
    };
```

We'll need a couple of extra includes in `Test.h` for `cstring` and `type_traits`, as follows:

```
#include <cstring>
#include <map>
#include <ostream>
#include <string_view>
#include <type_traits>
#include <vector>
```

The template specialization uses a new C++20 feature, called *requires*, which helps us to place constraints on template parameters. The `requires` keyword is actually part of a bigger enhancement in C++20, called *concepts*. Concepts are a huge enhancement to C++ and a full explanation would be beyond the scope of this book. We're using concepts and the `requires` keyword to simplify the template specialization to only work with strings. The template itself takes a T type like before and a new numeric value, N, which will be the size of the string literal. The requires clause makes sure that T is a char. We need to remove the const qualifier from T because string literals are actually constant.

The `Equals` specialization then says it is an array of T [N] . The constructor takes a reference to an array of N chars, and instead of trying to directly initialize mExpected with the constructor's `expected` parameter, it now calls memcpy to copy the chars from the literal into the mExpected array. The strange syntax of `char const (& expected) [N]` is how C++ specifies an array as a method parameter that does not get decayed into a simple pointer.

Now the `pass` method can take a string reference as its `actual` parameter type since we know that we are dealing with strings. Additionally, the `to_string` method can directly construct and return `std::string` from the mExpected char array.

One interesting, and maybe only theoretical, benefit of the `Equals` template specialization and the `pass` method is that we can now confirm that a string literal equals another string literal. I can't think of any place where this would be useful but it works, so we might as well add it to the test like this:

```
TEST("Test hamcrest style string confirms")
{
    std::string s1 = "abc";
    std::string s2 = "abc";
    CONFIRM_THAT(s1, Equals(s2));        // string vs. string
    CONFIRM_THAT(s1, Equals("abc"));     // string vs. literal
    CONFIRM_THAT("abc", Equals(s1));     // literal vs. string

    // Probably not needed, but this works too.
    CONFIRM_THAT("abc", Equals("abc")); // literal vs. Literal
}
```

What about char pointers? They're not as common as char arrays for template parameters because char arrays come from working with string literals. A char pointer is slightly different. We should consider char pointers because while they are not as common in template parameters, a char pointer is probably more common overall than char arrays. Here is a test that demonstrates char pointers. Note that this test will not compile yet. Add this into `Hamcrest.cpp`:

```
TEST("Test hamcrest style string pointer confirms")
{
    char const * sp1 = "abc";
    std::string s1 = "abc";
    char const * sp2 = s1.c_str();      // avoid sp1 and sp2
being same
    CONFIRM_THAT(sp1, Equals(sp2));     // pointer vs. pointer
    CONFIRM_THAT(sp2, Equals("abc"));  // pointer vs. literal
    CONFIRM_THAT("abc", Equals(sp2));  // literal vs. pointer
    CONFIRM_THAT(sp1, Equals(s1));     // pointer vs. string
    CONFIRM_THAT(s1, Equals(sp1));     // string vs. pointer
}
```

We can initialize a char pointer given a string literal just like how `std::string` is initialized. But while `std::string` copies the text into its own memory to manage, a char pointer just points to the first char in the string literal. I keep saying that we're working with char pointers. But to be more specific, we're working with constant char pointers. The code needs to use const, but I sometimes leave const out when speaking or writing.

The new test for the string pointer confirms the need to take extra steps to make sure that sp1 and sp2 point to different memory addresses.

String literals in C++ are consolidated so that duplicate literal values all point to the same memory address. Even though a literal such as "abc" might be used many times in the source code, there will only be one copy of the string literal in the final executable that gets built. The test must go through extra steps to make sure that sp1 and sp2 have different pointer values while maintaining the same text. Whenever std::string is initialized with a string literal, the text of the string literal gets copied into std::string to manage. The std::string might use dynamically allocated memory or local memory on the stack. A std::string will not just point to the memory address used in the initialization. If we simply initialized sp2 the same way as sp1, then both pointers would point to the same memory address. But by initializing sp2 to point to the string inside s1, then sp2 points to a different memory address from sp1. Even though sp1 and sp2 point to different memory addresses, the value of the text chars at each address is the same.

Okay, now that you understand what the test is doing, does it compile? No. The build fails while trying to call the pass method in the confirm_that template function.

The line in the test that causes the build failure is the last confirmation. The compiler is trying to convert the s1 string into a constant char pointer. But this is misleading because even if we comment out the last confirmation so that the build succeeds, the test then fails at runtime, like this:

```
------- Test: Test hamcrest style string pointer confirms
Failed confirm on line 75
    Expected: abc
    Actual   : abc
```

Because you might get different line numbers, I'll explain that line 75 is the first confirmation from the test:

```
CONFIRM_THAT(sp1, Equals(sp2));   // pointer vs. pointer
```

Look at the test failure message though. It says that "abc" is not equal to "abc"! What is going on?

Because we're using the original Equals template class, it only knows that we are dealing with char pointers. When we call pass, it's the pointer values that are being compared. And because we took extra steps to make sure that sp1 and sp2 have different pointer values, the test fails. And the test fails even though the text both pointers refer to is the same.

In order to support pointers, we'll need another template specialization of Equals. But we can't just specialize on any pointer type, in the same way we couldn't specialize on any array type. We made sure that the array specialization only works for char arrays. So, we should also make sure that our pointer specialization only works with char pointers. Add this specialization right after the second Equals class in Test.h:

```
template <typename T> requires (
    std::is_same<char, std::remove_const_t<T>>::value)
class Equals<T *> : public Matcher
{
public:
    Equals (char const * expected)
    : mExpected(expected)
    { }

    bool pass (std::string const & actual) const
    {
        return actual == mExpected;
    }

    std::string to_string () const override
    {
        return mExpected;
    }

private:
    std::string mExpected;
};
```

With this third version of the Equals class, we not only fix the build error but all the confirmations pass too! This template specializes Equals for T * and also requires that T be a char type.

The constructor accepts a pointer to constant chars and initializes mExpected with the pointer. The mExpected data member is std::string, which knows how to initialize itself from a pointer.

The pass method also accepts std::string, which will let it compare against actual strings or char pointers. Additionally, the to_string method can return mExpected directly since it's already a string.

When we were adding more classical confirmations in *Chapter 5, Adding More Confirm Types*, we added special support for floating point types. We'll need to add special support for confirming floating-point types in the Hamcrest style, too. The Hamcrest floating-point specializations will come in the next chapter along with learning how to write custom matchers.

Summary

We used TDD throughout this chapter to add Hamcrest confirmations and even improve the existing code for classical confirmations. Without TDD, the existing code in a real project would likely not get approval from management to make changes.

This chapter showed you the benefits of having unit tests that can help verify the quality of code after making changes. We were able to refactor the existing classical confirmations design for dealing with strings so that it matches the new design, which has a similar need. This lets both the classical and Hamcrest confirmations share a similar design instead of maintaining two different designs. All the changes were possible because the unit tests verified that everything continued to run as expected.

The most important changes in this chapter added Hamcrest style confirmations, which are more intuitive and more flexible than the classic confirmations developed in *Chapter 4, Adding Tests to a Project*. Additionally, the new Hamcrest confirmations are extensible.

We added support for Hamcrest confirmations following a TDD approach, which let us start simply. The simplicity was critical because we soon got into more advanced template specializations and even a new C++20 feature, called *requires*, that lets us specify how the templates should be used.

TDD makes the process of designing software flow better – from simple ideas at the start of a project or the beginning of an enhancement to an enhanced solution like this chapter developed. Even though we have working Hamcrest confirmations, we're not done yet. We'll continue to enhance the confirmations in the next chapter by making sure we can confirm floating-point values and custom-type values.

13

How to Test Floating-Point and Custom Values

We first encountered the need to test floating-point values in *Chapter 5, Adding More Confirm Types,* and created a simple solution that would let us compare floating-point values within a margin of error. We need the small margin because floating-point values that are close and might even look identical when displayed are almost always not exactly equal. These small differences make it hard to verify test results.

The main topics in this chapter are as follows:

- More precise floating-point comparisons
- Adding floating-point Hamcrest matchers
- Writing custom Hamcrest matchers

We're going to improve the simple solution developed earlier into a much better way to compare floating-point values that is more precise and works for both small and big values. We'll use the better comparison for both the earlier classical-style confirmations and the new Hamcrest-style confirmations.

You'll also learn how to create your own Hamcrest matchers in this chapter. We'll be creating a new matcher to test for inequality instead of always testing for equality, and you'll see how to contain one matcher inside another so that you can better reuse a matcher without needing to duplicate all the matcher template specializations.

Finally, you'll learn how to create another custom simple matcher that will be slightly different than the other matchers so far because the new matcher will not need an expected value.

Technical requirements

All code in this chapter uses standard C++ that builds on any modern C++ 20 or later compiler and standard library. The code is based on and continues enhancing the testing library from *Part 1* of this book, *Testing MVP*.

You can find all the code for this chapter in the following GitHub repository:

`https://github.com/PacktPublishing/Test-Driven-Development-with-CPP`

More precise floating-point comparisons

Whenever improvements are needed, one of the first things to look for is a way to measure the current design. Back in *Chapter 5*, *Adding More Confirm Types*, we examined floating-point numbers and I explained that comparing any floating-point type value—float, double, or long double—directly with another floating-point value is a bad idea. The comparison is too sensitive to small rounding errors and will usually result in the two values comparing not equal.

In *Chapter 5*, I showed you how to add a small margin to the comparison so that an accumulation of errors would not throw off the comparison as long as the two numbers being compared were close enough in value to each other. In other words, two values can compare equal as long as they are close enough to each other.

But what margin should be used? We simply picked some small numbers, and that solution worked. We're going to improve that solution. And now that you're becoming familiar with **test-driven development** (TDD), we're going to create some functions to help test our solution. All these functions will go at the top of `Hamcrest.cpp`.

The first function will convert a floating-point value into a fraction by dividing by a constant. We'll divide by 10, like this:

```
template <typename T>
T calculateFraction (T input)
{
    T denominator {10};
    return input / denominator;
}
```

This is a template, so it will work for float, double, and long double types. The intent is for the input to be a whole number, and this function will turn the number into tenths. Remember from *Chapter 5* that tenths don't have exact representations in binary. There will be a slight error introduced but not much because we only do one division calculation.

We'll need another function that will generate bigger margins of error by doing more work, like this function:

```
template <typename T>
T accumulateError (T input)
{
    // First add many small amounts.
    T partialAmount {0.1};
    for (int i = 0; i < 10; ++i)
    {
        input += partialAmount;
    }
    // Then subtract to get back to the original.
    T wholeAmount {1};
    input -= wholeAmount;
    return input;
}
```

This function adds 1 and then subtracts 1, so the input should remain unchanged. But because we add many small amounts that will all equal 1, the function introduces many errors during the calculations. The result that gets returned should be close to the original `input` but not the same.

The last helper function will call the first two functions many times for many different values and count how many times the results are equal. The function looks like this:

```
template <typename T>
int performComparisons (int totalCount)
{
    int passCount {0};
    for (int i = 0; i < totalCount; ++i)
    {
        T expected = static_cast<T>(i);
        expected = calculateFraction(expected);
        T actual = accumulateError(expected);
        if (actual == expected)
        {
            ++passCount;
        }
    }
}
```

```
        return passCount;
}
```

The function uses the fraction as the `expected` value since it should have the fewest errors. The `expected` value is compared with the `actual` value, which we get from accumulating many small errors. The two values should be close but not exactly equal. They should be close enough to be counted as equal, though.

Who defines what is close enough? That's really up to you to decide. The tests we're creating in this book might be allowing more errors than your application can tolerate. You'll understand after reading this section how to modify your code if you need more or less tolerance. There is no right answer for how to compare floating-point values that will work for all applications. The best you can do is to be aware of your own needs and adapt the code to fit those needs.

The `performComparisons` function also uses the `==` operator without any type of margin. The results should have a lot of unequal results. But how many? Let's write a test to find out!

Add this test to the end of `Hamcrest.cpp`:

```
TEST("Test many float comparisons")
{
    int totalCount {1'000};
    int passCount = performComparisons<float>(totalCount);
    CONFIRM_THAT(passCount, Equals(totalCount));
}
```

The test will cycle through `1,000` values, turning each one into tenths, introducing errors, and counting how many compared equal. The results are really bad:

```
------- Test: Test many float comparisons
Failed confirm on line 125
    Expected: 1000
    Actual  : 4
```

Only four values were close enough to be considered equal with the standard equality operator. You might get slightly different results depending on your computer and your compiler. If you do get different results, then that should be more evidence about how unreliable floating-point comparisons are. How about double and long double types? Add these two tests to find out:

```
TEST("Test many double comparisons")
{
    int totalCount {1'000};
```

```
    int passCount = performComparisons<double>(totalCount);
    CONFIRM_THAT(passCount, Equals(totalCount));
}


TEST("Test many long double comparisons")
{
    int totalCount {1'000};
    int passCount = performComparisons<long
                    double>(totalCount);
    CONFIRM_THAT(passCount, Equals(totalCount));
}
```

The results are just as bad and look like this:

```
------- Test: Test many double comparisons
Failed confirm on line 132
    Expected: 1000
    Actual  : 4
------- Test: Test many long double comparisons
Failed confirm on line 139
    Expected: 1000
    Actual  : 0
```

Let's add a margin to the equality comparison to see how much better the comparisons become. We'll start with the values used in the existing confirm overloads in Test.h. One of the overloads looks like this:

```
inline void confirm (
    float expected,
    float actual,
    int line)
{
    if (actual < (expected - 0.0001f) ||
        actual > (expected + 0.0001f))
    {
        throw ActualConfirmException(
            std::to_string(expected),
            std::to_string(actual),
```

```
                line);
        }
   }
```

The value we are interested in is the hardcoded literal floating-point value. In this case, it's 0.0001f. All we need to do is create three more helper functions that return these values. Note that the double and long double overloads have a different value than the float type. Place these three helper functions in Hamcrest.cpp, right before the performComparisons function, like this:

```
constexpr float getMargin (float)
{
    return 0.0001f;
}


constexpr double getMargin (double)
{
    return 0.000001;
}


constexpr long double getMargin (long double)
{
    return 0.000001L;
}
```

These three helper functions will let us customize the margin for each type. They each take a floating-point type parameter that is only used to determine which function to call. We don't actually need or use the parameter value passed to the function. We'll call these helper functions from within the performComparisons template, which will know the type to be used based on how the template was constructed.

We're also going to slightly change how we compare with a margin. Here's an example of how the confirm functions compare:

```
    if (actual < (expected - 0.0001f) ||
        actual > (expected + 0.0001f))
```

Instead of this, we're going to subtract the expected value from the actual value and then compare the absolute value of that subtraction result with the margin. We need to include cmath at the top of Hamcrest.cpp for the abs function, and we're going to need limits soon also, like this:

```
#include "../Test.h"

#include <cmath>
#include <limits>
```

And now, we can change the `performComparisons` function to use the margin, like this:

```
template <typename T>
int performComparisons (int totalCount)
{
    int passCount {0};
    for (int i = 0; i < totalCount; ++i)
    {
        T expected = static_cast<T>(i);
        expected = calculateFraction(expected);
        T actual = accumulateError(expected);
        if (std::abs(actual - expected) < getMargin(actual))
        {
            ++passCount;
        }
    }
    return passCount;
}
```

After making these changes, all the tests pass, like this:

```
------- Test: Test many float comparisons
Passed
------- Test: Test many double comparisons
Passed
------- Test: Test many long double comparisons
Passed
```

This means that all 1,000 values are now matching within a small margin of error. This is the same solution explained in *Chapter 5*. We should be good, right? Not quite.

The problem is that the margin value is quite big for small numbers and too small for big numbers. All the tests are passing, but that's just because we have a margin that's big enough to let a lot of comparisons be treated as equal.

To see this, let's refactor the comparison out of the `performComparisons` function so that the check is in its own function, like this:

```
template <typename T>
bool compareEq (T lhs, T rhs)
{
    return std::abs(lhs - rhs) < getMargin(lhs);
}


template <typename T>
int performComparisons (int totalCount)
{
    int passCount {0};
    for (int i = 0; i < totalCount; ++i)
    {
        T expected = static_cast<T>(i);
        expected = calculateFraction(expected);
        T actual = accumulateError(expected);
        if (compareEq(actual, expected))
        {
            ++passCount;
        }
    }
    return passCount;
}
```

And then we can write a couple tests to call `compareEq` directly, like this:

```
TEST("Test small float values")
{
    // Based on float epsilon = 1.1920928955078125e-07
    bool result = compareEq(0.000001f, 0.000002f);
    CONFIRM_FALSE(result);
}


TEST("Test large float values")
{
    // Based on float epsilon = 1.1920928955078125e-07
```

```
    bool result = compareEq(9'999.0f, 9'999.001f);
    CONFIRM_TRUE(result);
}
```

The test for small float values compares two numbers that are obviously different, yet the comparison function will consider them equal and the test fails. The fixed margin considers any float values within `0.0001f` to be equal. We want the two values to compare not equal, but our margin is big enough that they are considered to be equal.

What is the *epsilon* value that the comment refers to? We'll start using the actual epsilon values in just a moment, and this is why I suggested that you include `limits`. Floating-point numbers have a concept called epsilon, which is a value defined in `limits` for each floating-point type. The epsilon value represents the smallest distance between adjacent floating-point values for values between 1.0 and 2.0. Remember that floating-point values can't represent every possible fractional number, so there are gaps between the numbers that can be represented.

You can see the same thing yourself if you write down numbers with only a fixed number of decimal places on paper. Let's say that you limit yourself to only using two digits after the decimal point. You could write `1.00` and `1.01` and `1.02`. Those are adjacent values. In fact, `1.00` and `1.02` are the closest numbers you can represent to `1.01` by using only two digits after the decimal place. What about a number such as `1.011`? It's definitely closer to `1.01` than `1.02` but we can't write `1.011` because it needs three digits after the decimal point. The epsilon value for our experiment is `0.01`. Floating-point numbers have a similar problem except that the value of epsilon is smaller and not a simple value such as `0.01`.

Another complication is that the distance between adjacent floating-point numbers increases as the numbers get larger, and the distance decreases as the numbers get smaller. The test for small float values uses small values, but the values are much bigger than epsilon. Because the numbers are much bigger than epsilon, we want the test to fail. The test passes because our fixed margin is even bigger than epsilon.

The test for large float values also fails. It uses two values that are different by `0.001f`, which would be a really big difference if we were comparing `1.0f` with `1.001f`. At small values, a difference of `0.001f` would be enough to cause the values to compare not equal. But we're not dealing with small values—we're dealing with values that are almost 10,000! We now want the larger values to be considered equal because the fractional part makes up a smaller percentage of the larger numbers. The test fails because our fixed margin doesn't consider that the values are larger and only looks at the difference, which is greater than the fixed margin allows.

We can also test the other floating-point types. Add these two similar tests for double right after the two tests added for small and large float values, like this:

```
TEST("Test small double values")
{
```

```
    // Based on double epsilon = 2.2204460492503130808e-16
    bool result = compareEq(0.000000000000001,
                 0.000000000000002);
    CONFIRM_FALSE(result);
}


TEST("Test large double values")
{
    // Based on double epsilon = 2.2204460492503130808e-16
    bool result = compareEq(1'500'000'000'000.0,
                 1'500'000'000'000.0003);
    CONFIRM_TRUE(result);
}
```

For the double type, we have a different epsilon value that's much smaller than the epsilon for floats, and we also have many more significant digits to work with, so we can use numbers with more digits. We were limited to only about 7 digits when working with floats. With doubles, we can use numbers with about 16 digits. Notice that with doubles, we need a large value in the trillions in order to see a difference of 0.0003 that should be compared as equal.

If you are wondering how I arrived at these test numbers, I just picked small numbers just one decimal place bigger than epsilon for the small value tests. And for the large values, I choose a bigger number that I multiplied by (1 + epsilon) to arrive at the other number to be compared with. I then rounded the other number a bit so that it would be a bit closer. I had to choose a big number to start with that would stay within the number of digits allowed for each type.

Since we're using the double epsilon value for long doubles, the tests for small and large long doubles look similar to the tests for double. The long double tests look like this:

```
    TEST("Test small long double values")
    {
        // Based on double epsilon = 2.2204460492503130808e-16
        bool result = compareEq(0.000000000000001L,
                     0.000000000000002L);
        CONFIRM_FALSE(result);
    }


    TEST("Test large long double values")
    {
        // Based on double epsilon = 2.2204460492503130808e-16
```

```
    bool result = compareEq(1'500'000'000'000.0L,
                    1'500'000'000'000.0003L);
    CONFIRM_TRUE(result);
}
```

The only difference between the double tests and the long double tests is the long double suffix L at the end of the long double literal values.

After adding all six tests for small and large floating-point type tests, they all fail when run.

The reason for the failures is the same for each type. The small value tests all fail because the fixed margin considers the values to be equal when they should not be equal, and the large value tests consider the values to be not equal when they really are close, considering the large value. In fact, the large values are all within a single epsilon value from each other. The large values are as close as they can possibly get without being exactly equal. Sure—the long double large values could have been closer, but we're simplifying long doubles a bit by using the bigger epsilon from the double type.

We need to enhance the compareEq function so that the margin can be smaller for small values and bigger for big values. The moment we take on the responsibilities of comparing floating-point values, there are a lot of details that need to be handled. We skipped the extra details in *Chapter 5*. We're also going to skip some of the details even here. If you haven't realized it yet, dealing with floating-point values is really hard. The moment you think everything is working is when another detail comes along that changes everything.

Let's first fix the getMargin functions to return the real epsilon values modified slightly for each type, like this:

```
constexpr float getMargin (float)
{
    // 4 is chosen to pass a reasonable amount of error.
    return std::numeric_limits<float>::epsilon() * 4;
}


constexpr double getMargin (double)
{
    // 4 is chosen to pass a reasonable amount of error.
    return std::numeric_limits<double>::epsilon() * 4;
}


constexpr long double getMargin (long double)
{
    // Use double epsilon instead of long double epsilon.
```

```
    // Double epsilon is already much bigger than
    // long double epsilon so we don't need to multiply it.
    return std::numeric_limits<double>::epsilon();
}
```

The getMargin functions now use the epsilon values for the types as defined in numeric_limits. The margins are tuned for our needs. You might want to multiply by different numbers, and you might want to use the real epsilon value for long doubles. The reason we want bigger margins than epsilon itself is that we want to consider values to be equal that are more than just one epsilon value greater or lesser away from each other. We want a little more room for the accumulation of at least a few calculation errors. We multiply epsilon by 4 to give that extra room, and we use the double epsilon for long doubles, which might really be too much already. But these margins work for what we need.

We'll use the more accurate margin values in the new compareEq function, which looks like this:

```
template <typename T>
bool compareEq (T lhs, T rhs)
{
    // Check for an exact match with operator == first.
    if (lhs == rhs)
    {
        return true;
    }

    // Subnormal diffs near zero are treated as equal.
    T diff = std::abs(lhs - rhs);
    if (diff <= std::numeric_limits<T>::min())
    {
        return true;
    }

    // The margin should get bigger with bigger absolute
values.
    // We scale the margin up by the larger value or
    // leave the margin unchanged if larger is less than 1.
    lhs = std::abs(lhs);
    rhs = std::abs(rhs);
    T larger = (lhs > rhs) ? lhs : rhs;
```

```
        larger = (larger < 1.0) ? 1.0 : larger;
        return diff <= getMargin(lhs) * larger;
}
```

I like to use parameter names `lhs` and `rhs` for operator-type functions such as this. The abbreviations stand for left-hand side and right-hand side, respectively.

Consider these two numbers:

```
3 == 4
```

When making a comparison of these, 3 is on the left side of the operator and would be the `lhs` argument, while 4 is on the right and would be the `rhs` argument.

There's always a chance that the two numbers being compared are exactly equal to each other. So, the first thing we check is for an exact match using the `==` operator.

The `compareEq` function goes on to check the difference between the two numbers for a *subnormal* result. Remember I said that floating-point numbers are complicated? There could be an entire book written about floating-point math, and there are probably several books already written. I won't go into much explanation about subnormal values except to say that this is how floating-point values are represented when they are extremely close to zero. We'll consider any two subnormal values to be equal.

Subnormal values are also a good reason to *compare your numbers with each other* instead of *comparing their difference with zero*. You might wonder what the problem is. Doesn't the code in the `compareEq` function subtract one value from the other to arrive at the difference? Yes, it does. But our `compareEq` function doesn't try to compare the difference with zero directly. We figure out which of the two values is bigger and then scale the margin by multiplying the margin with the bigger value. We also avoid scaling the margin down when we are comparing values less than `1.0`.

If you have two values to compare and instead of passing them to `compareEq`, you pass their difference and compare the difference with zero, then you remove the ability of the `compareEq` function to do the scaling because the `compareEq` function would then only see a small difference and zero being compared.

The lesson here is to always pass your numbers to be compared directly to the `compareEq` function and let it figure out how much the two numbers are different by taking into account how big the numbers are. You'll get more accurate comparisons.

We could make the `compareEq` function even more elaborate. Maybe we could consider the sign of subnormal values instead of considering all of them to be equal, or maybe we could scale the margin down more so that we get very precise when dealing with subnormal values. This is not a book about math, so we're going to stop adding more to the `compareEq` function.

After making the changes to `compareEq`, all the tests pass. We now have a solution that allows small amounts of accumulated errors and lets two numbers compare equal when they are close enough.

The solution works for both really small numbers and really big numbers. The next section will turn the code we explored here into a better Hamcrest equality matcher.

Adding floating-point Hamcrest matchers

We explored better floating-point comparisons in the previous section, and now it's time to use the comparison code in the unit test library. Some of the code should be moved into `Test.h` where it fits better and can then be used by the test library. The rest of the code that was written should stay in `Hamcrest.cpp` because it's code that supports the tests.

The code that needs to be moved is the `compareEq` function and the three `getMargin` functions that `compareEq` calls to get the margins. We also need to move the includes of `cmath` and `limits` into `Test.h`, like this:

```
#include <cmath>
#include <cstring>
#include <limits>
#include <map>
#include <ostream>
#include <string_view>
#include <type_traits>
#include <vector>
```

The three `getMargin` functions and the `compareEq` function can be moved into `Test.h`, right before the first override of the `confirm` function that accepts Booleans. None of the code in the moved functions needs to change. Just cut the includes and the functions out of `Hamcrest.cpp` and paste the code into `Test.h`.

We might as well fix the existing floating-point classic `confirm` functions. This is why I had you move the `compareEq` function into `Test.h` immediately before the first `confirm` function. The change to the existing floating-point `confirm` functions is simple. They need to call `compareEq` instead of using hardcoded margins that don't scale. The `confirm` function for float types looks like this after the change:

```
inline void confirm (
    float expected,
    float actual,
    int line)
{
    if (not compareEq(actual, expected))
    {
```

```
        throw ActualConfirmException(
            std::to_string(expected),
            std::to_string(actual),
            line);
    }
}
```

The other two confirm functions that accept double and long double types should be changed to look similar. All three confirm functions will create the correct template of compareEq based on the expected and actual parameter types.

We should build and run the test application to make sure nothing broke with this small refactoring. And all the tests pass. We now have updated classic style confirm functions that will work better with floating-point comparisons.

We can make the code slightly better, though. We have three almost identical functions that are different only by their parameter types. The only reason for the three functions is that we want to override the confirm function for floating-point types. But since we're using C++20, let's use *concepts* instead! Concepts are a new feature that we've already started using when we specialized the Equals matcher to work with char arrays and char pointers in the previous chapter. Concepts allow us to tell the compiler which types are acceptable for template parameters and even function parameters. In the previous chapter, we were only using the requires keyword to place some restrictions on the template parameters. We'll be using more well-known concepts here in this chapter.

We need to include concepts like this in Test.h:

```
#include <concepts>
#include <cmath>
#include <cstring>
#include <limits>
#include <map>
#include <ostream>
#include <string_view>
#include <type_traits>
#include <vector>
```

And then, we can replace the three confirm functions that accept float, double, and long double types with a single template that uses the floating_point concept, like this:

```
template <std::floating_point T>
void confirm (
    T expected,
```

```
        T actual,
        int line)
    {
        if (not compareEq(actual, expected))
        {
            throw ActualConfirmException(
                std::to_string(expected),
                std::to_string(actual),
                line);
        }
    }
}
```

This new template will only accept floating-point types, and by making both expected and actual share the same type T, then both types must be the same. The definition of floating_point is one of the well-known concepts defined in the concepts header.

Now that we have the classic style confirmations working, let's get the Hamcrest Equals matcher working for floating-point values. We can first change the three large floating-point tests in Hamcrest. cpp to stop calling compareEq directly and instead use the CONFIRM_THAT macro so that they look like this:

```
TEST("Test large float values")
{
    // Based on float epsilon = 1.1920928955078125e-07
    CONFIRM_THAT(9'999.0f, Equals(9'999.001f));
}


TEST("Test large double values")
{
    // Based on double epsilon = 2.2204460492503130808e-16
    CONFIRM_THAT(1'500'000'000'000.0,
                 Equals(1'500'000'000'000.0003));
}


TEST("Test large long double values")
{
    // Based on double epsilon = 2.2204460492503130808e-16
    CONFIRM_THAT(1'500'000'000'000.0L,
                 Equals(1'500'000'000'000.0003L));
}
```

We're not going to change the tests for the small floating-point values yet because we don't have a matcher that does inequality comparisons. The solution might be as simple as putting the not keyword in front of Equals, but let's hold off on that for just a moment because we'll be exploring our options in the next section.

With the change to the tests, they should fail because we haven't yet specialized the Equals matcher to do anything different for floating-point types. Building and running the test application shows that the three tests do fail, like this:

```
------- Test: Test large float values
Failed confirm on line 152
    Expected: 9999.000977
    Actual  : 9999.000000
------- Test: Test small double values
Passed
------- Test: Test large double values
Failed confirm on line 165
    Expected: 1500000000000.000244
    Actual  : 1500000000000.000000
------- Test: Test small long double values
Passed
------- Test: Test large long double values
Failed confirm on line 178
    Expected: 1500000000000.000300
    Actual  : 1500000000000.000000
```

Notice how the expected values printed in the summary report don't exactly match the literal values given in the tests for the float and the double types. The long double does display a value that matches the value given in the test. The discrepancy is because floating-point variables are unable to always match exact values. The differences become more visible with floats, a little less visible with doubles, and closer to the desired values with long doubles.

The steps we just went through follow TDD. We modified existing tests instead of creating new tests because we don't expect callers to use compareEq directly. The tests were initially written to call compareEq directly to show that we had a solution for floating-point types that worked. Modifying the tests to the desired usage is the right thing to do, and then, by running the tests, we can see the failures. This is good because we expected the tests to fail. Had the tests passed instead, then we would need to find the reason for the unexpected success.

Let's get the tests to pass again! We need a version of Equals that knows how to work with floating-point types. We'll use the concept floating_point that we just used for the classic style confirmations to create another version of Equals that will call compareEq for floating-point types. Place this new Equals specialization in Test.h right after Equals that works with char pointers, like this:

```
template <std::floating_point T>
class Equals<T> : public Matcher
{
public:
    Equals (T const & expected)
    : mExpected(expected)
    { }

    bool pass (T const & actual) const
    {
        return compareEq(actual, mExpected);
    }

    std::string to_string () const override
    {
        return std::to_string(mExpected);
    }

private:
    T mExpected;
};
```

That's all we need to change to get the tests passing again. The new Equals specialization accepts any floating-point type and will be preferred by the compiler for floating-point types instead of the general-purpose Equals template. The floating-point version of Equals calls compareEq to do the comparison. We also don't need to worry about which types will be passed to to_string and can simply call std::to_string since we know that we will have one of the built-in floating-point types. The to_string assumption could fail if the user passes in some other type that has been created to be a floating_point concept type, but let's keep the code as simple as it can be for now and not worry about custom floating-point types.

The next section will start by creating a solution to test for inequality instead. We'll use the solution we create in the next section to modify the small floating-point Hamcrest tests.

Writing custom Hamcrest matchers

The previous section ended with the `Equals` matcher specialized to call `compareEq` for floating-point types. We also modified the tests for large floating-point values because they could use the Hamcrest style and the `Equals` matcher.

We left the tests for small floating-point values unchanged because those tests need to make sure that the actual and expected values are not equal.

We want to update the small floating-point value tests and need a way to test for not equal values. Maybe we could create a new matcher called `NotEquals`, or we could put the `not` keyword in front of the `Equals` matcher.

I'd like to avoid the need for a new matcher if possible. We don't really need any new behavior—we just need to flip the results of the existing `Equals` matcher. Let's try modifying the small floating-point value tests to look like this in `Hamcrest.cpp`:

```
TEST("Test small float values")
{
    // Based on float epsilon = 1.1920928955078125e-07
    CONFIRM_THAT(0.000001f, not Equals(0.000002f));
}


TEST("Test small double values")
{
    // Based on double epsilon = 2.2204460492503130808e-16
    CONFIRM_THAT(0.000000000000001,
            not Equals(0.000000000000002));
}


TEST("Test small long double values")
{
    // Based on double epsilon = 2.2204460492503130808e-16
    CONFIRM_THAT(0.000000000000001L,
            not Equals(0.000000000000002L));
}
```

The only changes are to stop calling `compareEq` directly and to use the `CONFIRM_THAT` macro with the `Equals` matcher. Notice that we flip the results of the `Equals` matcher by placing the `not` keyword in front.

Does it build? No. We get compile errors similar to this:

```
MereTDD/tests/Hamcrest.cpp:145:29: error: no match for
'operator!' (operand type is 'MereTDD::Equals<float>')
  145 |        CONFIRM_THAT(0.000001f, not Equals(0.000002f));
      |                                    ^~~~~~~~~~~~~~~~~~~
```

The not keyword in C++ is a shortcut for operator !. Normally with TDD, the next step would be to modify the code so that the tests can build. But we have a problem. The not keyword expects the class to have an operator ! method or some way to cast the class to a Boolean. Either option requires that the class be able to generate a bool value, and that's not how the matchers work. In order for a matcher to know whether the result should pass or not, it needs to know the actual value. The confirm_that function passes the matcher the needed actual value as an argument to the pass method. We can't turn a matcher by itself into a bool result.

We're going to have to create a NotEquals matcher. While not my first preference, a new matcher is acceptable from a test perspective. Let's change the tests to look like this instead:

```
TEST("Test small float values")
{
    // Based on float epsilon = 1.1920928955078125e-07
    CONFIRM_THAT(0.000001f, NotEquals(0.000002f));
}


TEST("Test small double values")
{
    // Based on double epsilon = 2.2204460492503130808e-16
    CONFIRM_THAT(0.000000000000001,
            NotEquals(0.000000000000002));
}


TEST("Test small long double values")
{
    // Based on double epsilon = 2.2204460492503130808e-16
    CONFIRM_THAT(0.000000000000001L,
            NotEquals(0.000000000000002L));
}
```

Another reason that I wanted to avoid a new matcher is to avoid the need to specialize the new matcher as we did for the Equals matcher, but there is a way to create a matcher called NotEquals and base its implementation on all the work we did for the Equals matcher. All we need to do is contain the Equals matcher and flip the pass result, like this:

```
template <typename T>
class NotEquals : public Matcher
{
public:
    NotEquals (T const & expected)
    : mExpected(expected)
    { }

    template <typename U>
    bool pass (U const & actual) const
    {
        return not mExpected.pass(actual);
    }

    std::string to_string () const override
    {
        return "not " + mExpected.to_string();
    }

private:
    Equals<T> mExpected;
};
```

Add the NotEquals matcher right after all the template specializations of the Equals matcher in Test.h.

The NotEquals matcher is a new matcher type that contains an Equals matcher for its mExpected data member. This will give us all the specialization we did for the Equals matcher. Whenever the NotEquals::pass method is called, we just call the mExpected.pass method and flip the result. And whenever the to_string method is called, we just add "not " to whichever string mExpected provides.

One interesting thing to notice is that the pass method is itself a template based on a type U. This will let us construct a NotEquals matcher given a string literal and then call the pass method with std::string.

We should add a test for using the NotEquals matcher with a string literal and std::string or, even better, to extend an existing test. We have two tests already that work with strings, string literals, and char pointers. Both tests are in Hamcrest.cpp. The first test should look like this:

```cpp
TEST("Test hamcrest style string confirms")
{
    std::string s1 = "abc";
    std::string s2 = "abc";
    CONFIRM_THAT(s1, Equals(s2));        // string vs. string
    CONFIRM_THAT(s1, Equals("abc"));     // string vs. literal
    CONFIRM_THAT("abc", Equals(s1));     // literal vs. string

    // Probably not needed, but this works too.
    CONFIRM_THAT("abc", Equals("abc")); // literal vs. literal

    std::string s3 = "def";
    CONFIRM_THAT(s1, NotEquals(s3));        // string vs. string
    CONFIRM_THAT(s1, NotEquals("def"));     // string vs.
literal
    CONFIRM_THAT("def", NotEquals(s1));     // literal vs.
string
}
```

And the second test should be modified to look like this:

```cpp
TEST("Test hamcrest style string pointer confirms")
{
    char const * sp1 = "abc";
    std::string s1 = "abc";
    char const * sp2 = s1.c_str();    // avoid sp1 and sp2
being same
    CONFIRM_THAT(sp1, Equals(sp2));   // pointer vs. pointer
    CONFIRM_THAT(sp2, Equals("abc")); // pointer vs. literal
    CONFIRM_THAT("abc", Equals(sp2)); // literal vs. pointer
    CONFIRM_THAT(sp1, Equals(s1));    // pointer vs. string
    CONFIRM_THAT(s1, Equals(sp1));    // string vs. pointer

    char const * sp3 = "def";
    CONFIRM_THAT(sp1, NotEquals(sp3));   // pointer vs. pointer
```

```
    CONFIRM_THAT(sp1, NotEquals("def")); // pointer vs. literal
    CONFIRM_THAT("def", NotEquals(sp1)); // literal vs. pointer
    CONFIRM_THAT(sp3, NotEquals(s1));    // pointer vs. string
    CONFIRM_THAT(s1, NotEquals(sp3));    // string vs. pointer
}
```

Building and running the test application shows that all the tests pass. Instead of adding new tests, we were able to modify the existing tests because the two existing tests were focused on the string and char pointer types. The NotEquals matcher fit right into the existing tests.

Having both Equals and NotEquals matchers gives us more than we had with the classic style confirmations, and we can go further by creating another matcher. You can also create matchers to do whatever you want in your test projects. We're going to create a new matcher in the MereTDD namespace but you can put yours in your own namespace. The matcher we'll be creating will test to make sure that an integral number is even. We'll call the matcher IsEven, and we can write a couple of tests in Hamcrest.cpp to look like this:

```
TEST("Test even integral value")
{
    CONFIRM_THAT(10, IsEven<int>());
}

TEST("Test even integral value confirm failure")
{
    CONFIRM_THAT(11, IsEven<int>());
}
```

You'll notice something different about the IsEven matcher: it doesn't require an expected value. The matcher only needs the actual value passed to it in order to confirm whether the actual value is even or not. Because there's nothing to pass to the constructor when creating an IsEven matcher in the test, we need to specify the type, like this:

```
IsEven<int>()
```

The second test should fail, and we'll use the failure to get the exact error message so that we can turn the test into an expected failure. But we first need to create an IsEven matcher. The IsEven class can go in Test.h immediately after the NotEquals matcher, like this:

```
template <std::integral T>
class IsEven : public Matcher
{
public:
```

```
        IsEven ()
        { }

        bool pass (T const & actual) const
        {
            return actual % 2 == 0;
        }

        std::string to_string () const override
        {
            return "is even";
        }
    };
```

I wanted to show you an example of a really simple custom matcher so that you'll know that they don't all need to be complicated or have multiple template specializations. The IsEven matcher just tests the actual value in the pass method to make sure it's even, and the to_string method returns a fixed string.

Building and running shows the even value test passes while the intended failure test fails, like this:

```
------- Test: Test even integral value
Passed
------- Test: Test even integral value confirm failure
Failed confirm on line 185
    Expected: is even
    Actual   : 11
```

With the error message, we can modify the even confirm failure test so that it will pass with an expected failure, like this:

```
TEST("Test even integral value confirm failure")
{
    std::string reason = "    Expected: is even\n";
    reason += "    Actual   : 11";
    setExpectedFailureReason(reason);

    CONFIRM_THAT(11, IsEven<int>());
}
```

Building and running now shows that both tests pass. One passes successfully and the other passes with an expected failure, like this:

```
------- Test: Test even integral value
Passed
------- Test: Test even integral value confirm failure
Expected failure
    Expected: is even
    Actual  : 11
```

That's all there is to making custom matchers! You can make matchers for your own classes or add custom matchers for new behaviors. Maybe you want to verify that a number has only a certain number of digits, that a string begins with some given text prefix, or that a log message contains a certain tag. Remember how in *Chapter 10, The TDD Process in Depth*, we had to verify tags by writing to a file and then scanning the file to make sure the line just written contained a tag? We could have a custom matcher that looks for a tag instead.

Summary

One of the main benefits of the Hamcrest style confirmations is their ability to be extended through custom matchers. What better way to explore this ability than through floating-point confirmations? Because there is no single best way to compare floating-point values, you might need a solution that's tuned to your specific needs. You learned about a good general-purpose floating-point comparison technique in this chapter that scales a small margin of error so that bigger floating-point values are allowed to differ by greater amounts as the values get bigger and still be considered to be equal.

If this general solution doesn't meet your needs, you now know how to create your own matcher that will do exactly what you need.

And the ability to extend matchers doesn't stop at floating-point values. You might have your own custom behavior that you need to confirm, and after reading this chapter, you now know how to create a custom matcher to do what you need.

Not all matchers need to be big and complicated and have multiple template specializations. You saw an example of a very simple custom matcher that confirms whether a number is even or not.

We also made good use of the concepts feature, new in C++20, which allows you to easily specify requirements on your template types. We made good use of concepts in this chapter to make sure that the floating-point matcher only works for floating-point types and that the `IsEven` matcher only works for integral types. You can use concepts in your matchers too, which will help you control how your matchers can be used.

The next chapter will explore how to test services and will introduce a new service project that uses all the code developed so far in this book.

14

How to Test Services

We've been building up to this point where we can use both the testing library and the logging library in another project. The customer of the logging library has always been a microservices C++ developer who is using TDD to design better services.

Because of the focus on services, this chapter will introduce a project that simulates a microservice. We're not going to include everything that a real service would need. For example, a real service needs networking and the ability to route and queue requests and handle timeouts. Our service will only contain the core methods to start the service and handle requests.

You'll learn about the challenges involved with testing services and how testing a service is different from testing an application that tries to do everything. There will be less focus in this chapter on the design of the service. And we're not going to be writing all the tests needed. In fact, this entire chapter only uses a single test. Other tests are mentioned that can be added.

We'll also explore what can be tested in a service, along with some tips and guidance that will enable you to control the amount of logging that gets generated when debugging a service.

The service project will help tie together the testing and logging libraries and show you how to use both libraries in your own projects.

The main topics in this chapter are as follows:

- Service testing challenges
- What can be tested in a service?
- Introducing the SimpleService project

Technical requirements

All code in this chapter uses standard C++ that builds on any modern C++ 20 or later compiler and standard library. The code introduces a new service project that uses the testing library from *Part 1, Testing MVP*, of this book, and uses the logging library from *Part 2, Logging Library*, of this book.

You can find all the code for this chapter in the following GitHub repository:

`https://github.com/PacktPublishing/Test-Driven-Development-with-CPP`

Service testing challenges

The customer we've been thinking about throughout this book has been a microservices developer writing services in C++ who wants to better understand TDD to improve the development process and increase the code quality. TDD is for anybody writing code. But in order to follow TDD, you need to have a clear idea of who your customer is so that you can write tests from that customer's viewpoint.

There are different challenges involved when testing services as compared to testing an application that does everything itself. An application that includes everything is often called a *monolithic application*. Some examples of challenges that apply to services are these:

- Is the service reachable?

- Is the service running?

- Is the service overloaded with other requests?

- Are there any permissions or security checks that could affect your ability to call a service?

Before we get too far though, we need to understand what a service is and why you should care.

A service runs on its own and receives requests, processes the requests, and returns some type of response for each request. A service is focused on the requests and responses, which makes them easier to write and debug. You don't have to worry about other code interacting with your service in unexpected ways because the request and response fully define the interaction. If your service starts getting too many requests, you can always add more instances of the service to handle the extra load. When services are focused on handling a few specific requests, they're called *microservices*. Building large and complicated solutions becomes easier and more reliable when you can divide the work into microservices.

Services can also make requests to other services in order to process a request. This is how microservices can build on each other to form bigger services. At each step, the request and expected response are clear and well defined. Maybe your entire solution is composed entirely of services. But more likely, you'll have an application that a customer runs, which accepts the customer's input and direction and makes requests from various services to fulfill the customer's need. Maybe the customer opens an application window that displays a chart of information based on some dates that the customer provides. In order to get the data to display the chart, the application will send the dates in a request to a service that will respond with the data. The service might even customize the data based on the specific customer making the request.

Imagine how much harder it would be to write an application that tried to do everything itself. The development effort might even go from an unthinkably complicated monolithic application to a reasonable effort when using services. The quality also goes up when tasks can be isolated and developed, and managed independently as services.

Services typically run on multiple computers, so the requests and responses are made over a network. There could be other routing code involved too that accepts a request and puts it in a queue before sending the request to a service. A service might be running on multiple computers, and the router will figure out which service is best able to process a request.

If you're lucky enough to have a large and well-designed network of services, then you'll probably have multiple separate networks designed to help you test and deploy your services. Each network can have many different computers, and each computer can be running multiple different services. This is where routers become very useful.

Testing a service that's running in multiple networks typically involves deploying a new version of the service to be tested on a computer in one of the networks designed for early testing. This network is usually called a *development environment*.

If the tests fail in the development environment, then you have time to find the bugs, make changes, and test a new version, until the service runs as expected. Finding bugs involves looking at the responses to make sure they are correct, examining the log files to make sure the correct steps were taken along the way, and looking at any other output, such as database entries that might have been modified while processing a request. Depending on the service, you might have other things to check.

Some services depend on data stored in databases to properly respond to requests. It might be difficult to keep the databases current in a development environment, which is why other environments are usually needed. If the initial tests pass in the development environment, then you might deploy the service changes to a *beta environment* and test again. Eventually, you'll deploy the service to the *production environment* where it will serve responses for customers.

If you can control the routing of requests, then it might be possible for you to run a debugger when testing your changes. The way to do this is to start the service on a particular computer under the debugger. Usually, this will only be done in the development environment. Then, you will need to make sure that any requests you make through a test user account get routed to the computer where you have the debugger running. The same service without your recent changes will likely be running on other computers in the same environment, which is why debugging with a debugger only works if you can make sure that the requests will be routed to the computer you're using.

If you don't have the ability to route requests to a specific computer or if you're testing in an environment that doesn't allow debuggers, then you'll need to rely heavily on the log messages. Sometimes you won't know ahead of time which computer in an environment will handle a request, so you'll need to deploy your service to all the computers in that environment.

Examining log files can be tedious because you need to visit each computer just to open the log files to see if your testing request was handled on that computer or some other computer. If you have a service that gathers log files from each computer and makes the log messages available for searching, then you'll have a much easier time testing your service in environments that have multiple computers.

You don't have the same distributed testing problems when testing a single application that doesn't use services. You can even use your own computer for much of the testing. You can run your changes under a debugger, examine log files, and run unit tests quickly and directly. Services require much more support, such as a message routing infrastructure that you might not be able to set up on your own computer.

Every company and organization that builds solutions with microservices will have different environments and deployment steps. There's no way that I can tell you how to test your particular service. And that's not the goal of this section. I'm only explaining the challenges with testing services that are different from testing an application that tries to do everything.

Even with all the extra networking and routing, services are a great way to design a large application. Who knows, the routing might even be a service itself. With all the isolated and independent services, it becomes possible to add new features and upgrade the user experience in small steps instead of releasing a new version that does everything.

Using services for a small application might not be worth the overhead. But I've seen a lot of small applications that grew into large applications and then got stuck when the complexity became too much. And the same thing happens with services and the language they are written in. I've seen services start out so small they could be written in a few lines of Python code. The developers might have been under a tight deadline and writing a small service in Python was faster than writing the same service in C++. Eventually, the small service proves to be valuable to other teams and grows in usage and in features. And it continues to grow until it needs to be replaced by a service written in C++.

Now that you know a bit more about the challenges of testing services, the next section explores what can be tested.

What can be tested in a service?

Services are defined by the requests accepted and the responses returned. A service will also have an address or some means of routing the requests to the service. There could be a version number or maybe the version is included as part of the address.

When putting all this together, you first need to prepare a request and send the request to the service. The response might come all at once or in pieces. Or maybe the response is like a ticket that you can present at a later time to the same or a different service to get the actual response.

All this means that there are different ways to interact with a service. About the only thing that remains the same is the basic idea of a request and then a response. If you make a request, there's no guarantee that a service will receive the request. And if a service replies, there's no guarantee that the response will make it back to the original requestor. Handling timeouts is always a big concern when working with services.

You probably won't want to test for timeouts directly because it can easily be anywhere from 30 seconds to 5 minutes before a service request is aborted due to no response. But you might want to test for response times within an expected and reasonable time. Be careful with tests like this though because they can sometimes pass and sometimes fail depending on many factors that can change and are outside of the direct control of the test. A timeout test is also more of a stress test or an acceptance test, and while it might help identify a poor design after the service has been deployed, focusing on timeouts initially is usually the wrong choice for TDD.

Instead, treat a service just like any other software that you'll be designing with TDD. Be clear about who the customer is and what their needs are, and then come up with a request and a response that makes the most sense, is easy to use, and is easy to understand.

When testing a service, it might be enough to make sure that the response contains the correct information. This will likely be the case for services that are completely out of your control. But it can be useful for a service to remain completely detached from any calling code and only interact through the request and the response.

Maybe you're calling a service that was created by a different company and the response is the only way to get the information requested. If so, then why are you testing the service? Remember to only test your code.

Assuming this is your service that you're designing and testing and that the response fully contains the information requested, then you can write tests that only need to form a request and examine the response.

Other times, a request might result in a response that simply acknowledges the request. The actual results of the request can appear elsewhere. In this case, you'll want to write tests that form a request, verify the response, and then verify the actual results wherever they are. Let's say you're designing a service that lets callers request that a file be deleted. The request would contain information about the file. The response might just be an acknowledgment that the file was deleted. And the test might then need to look in the folder where the file used to be located to make sure the file is no longer available.

Usually, requests that ask a service to do something will result in the need to verify that the action really was performed. And requests that ask a service to calculate something or return something might be able to confirm the information directly in the response. If the requested information is really big, then it might be better to find a different way to return the information.

However you design your service, the main point is that there are many options. You'll want to consider how your service will be used when writing your tests to create the design.

You might even need to call two or more services in your tests. For example, if you're writing a service that's designed to replace an older service with a slow calculation response time, you might want to call both services and compare the returned information to make sure that the new service is still returning the same information as the older service.

Services have a lot of overhead involved with the formatting and routing of the request, and the interpretation of the response. Testing a service is not like simply calling a function.

However, at some point internally, a service will contain a function to process or handle the request. This function is usually not exposed to users of the service. The users must go through the service interface, which involves routing a request and a response through a network connection.

Because C++ doesn't yet have standard networking capabilities, which might arrive in C++23, we're going to skip over all the networking and official request and response definitions. We'll create a simple service that resembles what a real service would look like internally.

We'll also focus on the type of service request that can return information completely in the response. The next section will introduce the service.

Introducing the SimpleService project

We're going to start a new project in this section to build a service. And just like how the logging project uses the testing project, this service project will use the testing project. The service will go further and also use the logging project. The service won't be a real service because a full service needs a lot of supporting code that is not standard C++ and would take us into topics unrelated to learning TDD.

The service will be called `SimpleService` and the initial set of files will tie together many of the topics already explained in this book. Here is the project structure:

```
SimpleService project root folder
    MereTDD folder
        Test.h
    MereMemo folder
        Log.h
    SimpleService folder
        tests folder
            main.cpp
            Message.cpp
            SetupTeardown.cpp
            SetupTeardown.h
        LogTags.h
        Service.cpp
        Service.h
```

When I started this project, I didn't know what files would be needed. I knew the project would use `MereTDD` and `MereMemo` and would have its own folder for the service. Inside the `SimpleService` folder, I knew there would be a `tests` folder that would contain `main.cpp`. I guessed there would be `Service.h` and `Service.cpp` too. I also added a file for the first test called `Message.cpp`. The idea of the first test would be something that would send a request and receive a response.

So let's start with the files that I knew would be in the project. `Test.h` and `Log.h` are the same files we've been developing so far in this book, and the `main.cpp` file looks similar, as follows:

```
#include <MereTDD/Test.h>

#include <iostream>

int main ()
{
    return MereTDD::runTests(std::cout);
}
```

The `main.cpp` file is actually a bit simpler than before. We're not using any default log tags so there's no need to include anything about logging. We just need to include the testing library and run the tests.

The first test that I wrote went in `Message.cpp` and looked like this:

```
#include "../Service.h"

#include <MereTDD/Test.h>

using namespace MereTDD;

TEST("Request can be sent and response received")
{
    std::string user = "123";
    std::string path = "";
    std::string request = "Hello";
    std::string expectedResponse = "Hi, " + user;

    SimpleService::Service service;
    service.start();

    std::string response = service.handleRequest(
```

```
                user, path, request);
       CONFIRM_THAT(response, Equals(expectedResponse));
}
```

My thinking at the time was there would be a class called `Service` that could be constructed and started. Once the service was started, we could call a method called `handleRequest`, which would need a user ID, a service path, and the request. The `handleRequest` method would return the response, which would be a string.

The request would also be a string and I decided to go with a simple greeting service. The request would be the `"Hello"` string and the response would be `"Hi, "` followed by the user ID. I put a Hamcrest-style confirmation of the response in the test.

I realized that we would eventually need other tests, and the other tests should use a service that was already started. Reusing an already running service would be better than creating an instance of the service and starting the service each time a test is run. So, I changed the `Message.cpp` file to use a test suite with setup and teardown like this:

```
#include "../Service.h"

#include "SetupTeardown.h"

#include <MereTDD/Test.h>

using namespace MereTDD;

TEST_SUITE("Request can be sent and response received",
"Service 1")
{
    std::string user = "123";
    std::string path = "";
    std::string request = "Hello";
    std::string expectedResponse = "Hi, " + user;

    std::string response = gService1.service().handleRequest(
        user, path, request);
    CONFIRM_THAT(response, Equals(expectedResponse));
}
```

This is the only test we're going to add to the service in this chapter. It will be enough to send a request and get a response.

I added the `SetupTeardown.h` and `SetupTeardown.cpp` files to the `tests` folder. The header file looks like this:

```cpp
#ifndef SIMPLESERVICE_TESTS_SUITES_H
#define SIMPLESERVICE_TESTS_SUITES_H

#include "../Service.h"

#include <MereMemo/Log.h>
#include <MereTDD/Test.h>

class ServiceSetup
{
public:
    void setup ()
    {
        mService.start();
    }

    void teardown ()
    {
    }

    SimpleService::Service & service ()
    {
        return mService;
    }

private:
    SimpleService::Service mService;
};

extern MereTDD::TestSuiteSetupAndTeardown<ServiceSetup>
gService1;

#endif // SIMPLESERVICE_TESTS_SUITES_H
```

This file contains nothing you haven't seen already in this book. Except that we previously declared setup and teardown classes in a single test .cpp file. This is the first time we've needed to declare setup and teardown in a header file so it can be reused in other test files later. You can see that the setup method calls the start method of the service. The only real difference is that the gService1 global instance needs to be declared extern, so we don't get linker errors later with other test files also using the same setup and teardown code.

The SetupTeardown.cpp file looks like this:

```
#include "SetupTeardown.h"

MereTDD::TestSuiteSetupAndTeardown<ServiceSetup>
gService1("Greeting Service", "Service 1");
```

This is simply the instance of gService1 that was declared extern in the header file. The suite name "Service 1" needs to match the suite name used in the TEST_SUITE macro in Message.cpp.

Moving on to the Service class declaration in Service.h, it looks like this:

```
#ifndef SIMPLESERVICE_SERVICE_H
#define SIMPLESERVICE_SERVICE_H

#include <string>

namespace SimpleService
{

class Service
{
public:
    void start ();

    std::string handleRequest (std::string const & user,
        std::string const & path,
        std::string const & request);
};

} // namespace SimpleService

#endif // SIMPLESERVICE_SERVICE_H
```

I put the service code in the `SimpleService` namespace, which you saw in the original test and in the setup and teardown code. The `start` method needs no parameters and returns void. At least for now, anyway. We can always enhance the service later. I felt it was important to include the idea of starting a service from the very beginning, even if there's not much to do yet. The idea that a service is already running and waiting to process requests is a core concept that defines what a service is.

The other method is the `handleRequest` method. We're skipping over a lot of details of a real service, such as the definition of requests and responses. A real service would have a documented way to define requests and responses, almost like a programming language itself. We're just going to use strings for both the request and the response.

A real service would use authentication and authorization to verify users and what each user is allowed to do with the service. We're simply going to use a string as the `user` identity.

And some services have an idea called a *service path*. The path is not the address of the service. The path is like a call stack in programming terms. Usually, the router would start the path whenever an application makes a call to a service. The `path` parameter acts like a unique identifier for the call itself. If the service needs to call other services in order to process the request, then the router for these additional service requests would add to the initial `path` that was already started. Each time `path` grows, the router adds another unique identifier to the end of the `path`. The `path` can be used in the service to log messages.

The whole point of the `path` is so that developers can make sense of the log messages by relating and ordering log messages for specific requests. Remember that a service is handling requests all the time from different users. And calling other services will cause those other services to log their own activity. Having a `path` that identifies a single service request and all of its related service calls, even across multiple log files, is really helpful when debugging.

The implementation of the service is in `Service.cpp` and looks like this:

```
#include "Service.h"

#include "LogTags.h"

#include <MereMemo/Log.h>

void SimpleService::Service::start ()
{
    MereMemo::FileOutput appFile ("logs");
    MereMemo::addLogOutput (appFile);

    MereMemo::log (info) << "Service is starting.";
```

```
    }

    std::string SimpleService::Service::handleRequest (
        std::string const & user,
        std::string const & path,
        std::string const & request)
    {

        MereMemo::log(debug, User(user), LogPath(path))
            << "Received: " << Request(request);

        std::string response;
        if (request == "Hello")
        {
            response = "Hi, " + user;
        }
        else
        {
            response = "Unrecognized request.";
        }

        MereMemo::log(debug, User(user), LogPath(path))
            << "Sending: " << Response(response);
        return response;
    }
```

Some books and guidance for TDD will say that this is too much code for a first test, that there should not be any logging or checking of the request string, and that, really, the first implementation should return an empty string just so that the test will fail.

Then the response should be hardcoded to be the exact value that the test expects. And then another test should be created that uses a different user ID. Only then should the response be built by looking at the user ID passed to the handleRequest method.

Checking the request against known values should come later after more tests are created that pass in different request strings. I'm sure you get the idea.

While I do like to follow steps, I think there's a balance more toward writing a little extra code so that the TDD process doesn't get too tedious. This initial service still does very little. And adding the logging and some of the initial structure to the code helps lay the foundation for what will come later. At least that's my opinion.

For the logging, you'll notice some things such as `User(user)` in the `log` calls. These are custom logging tags, like those we built in *Chapter 10, The TDD Process in Depth*. All the custom tags are defined in the last project file called `LogTags.h`, which looks like this:

```cpp
#ifndef SIMPLESERVICE_LOGTAGS_H
#define SIMPLESERVICE_LOGTAGS_H

#include <MereMemo/Log.h>

namespace SimpleService
{

inline MereMemo::LogLevel error("error");
inline MereMemo::LogLevel info("info");
inline MereMemo::LogLevel debug("debug");

class User : public MereMemo::StringTagType<User>
{
public:
    static constexpr char key[] = "user";

    User (std::string const & value,
        MereMemo::TagOperation operation =
            MereMemo::TagOperation::None)
    : StringTagType(value, operation)
    { }
};

class LogPath : public MereMemo::StringTagType<LogPath>
{
public:
    static constexpr char key[] = "logpath";

    LogPath (std::string const & value,
        MereMemo::TagOperation operation =
            MereMemo::TagOperation::None)
    : StringTagType(value, operation)
```

```
        { }
    };


    class Request : public MereMemo::StringTagType<Request>
    {
    public:
        static constexpr char key[] = "request";

        Request (std::string const & value,
            MereMemo::TagOperation operation =
                MereMemo::TagOperation::None)
        : StringTagType(value, operation)
        { }
    };


    class Response : public MereMemo::StringTagType<Response>
    {
    public:
        static constexpr char key[] = "response";

        Response (std::string const & value,
            MereMemo::TagOperation operation =
                MereMemo::TagOperation::None)
        : StringTagType(value, operation)
        { }
    };


} // namespace SimpleService


#endif // SIMPLESERVICE_LOGTAGS_H
```

This file defines the custom User, LogPath, Request, and Response tags. The named log levels, error, info, and debug, are also defined. All of the tags are placed in the same SimpleService namespace as the Service class.

Note that the logging project also included a file called LogTags.h, which was put in the tests folder because we were testing the logging itself. For this service project, the LogTags.h file is in the service folder because the tags are part of the service. We're no longer testing that tags work.

We're not even testing that logging works. The tags get logged as part of the normal service operation, so they are now part of the service project.

With everything in place, we can build and run the test project, which shows the single test is passing. The summary report actually shows three tests passing because of the setup and teardown. The report looks like this:

```
Running 1 test suites
--------------- Suite: Service 1
------- Setup: Service 1
Passed
------- Test: Message can be sent and received
Passed
------- Teardown: Service 1
Passed
----------------------------------
Tests passed: 3
Tests failed: 0
```

And we can also look at the log file, which contains these messages:

```
2022-08-14T05:58:13.543 log_level="info" Service is starting.
2022-08-14T05:58:13.545 log_level="debug" logpath="" user="123"
Received: request="Hello"
2022-08-14T05:58:13.545 log_level="debug" logpath="" user="123"
Sending: response="Hi, 123"
```

Now we can see the core structure that makes up a service. The service is first started and ready to handle requests. When a request arrives, the request is logged, the processing takes place to produce a response, and then the response is logged before sending the response back to the caller.

We're using the testing library to simulate a real service by skipping over all the networking and routing and going straight to the service to start the service and handle requests.

We're not going to add any more tests at this time. But for your own service projects, that would be your next step. You would add a test for each request type if your service supports multiple different requests. And don't forget to add a test for an unrecognized request.

Each request type might have multiple tests for different combinations of request parameters. Remember that a real request in a real service will have the ability to define rich and complex requests where the request can specify its own set of parameters, just like how a function can define its own parameters.

Each request type will usually have its own response type. And you might have a common response type for errors. Either that or each response type will need to include fields for error information. It's probably easier if your response types are used for successful responses and any error responses return a standard error response type that you define.

Another good idea when testing services is to create a logging tag for each request type. We only have a single greeting request but imagine a service with several different requests that can be handled. If each log message was tagged with the request type, then it becomes easy to enable debug logging for just one type of request.

Right now, we're tagging the log messages with the user ID. This is another great way to enable debug-level logging without flooding the log file with too many log messages. We can set a filter to log debug log entries for a specific test user ID. We would also need a default filter set to `info`. We can then combine the user ID with the request type to get even more precise. Once the filters are set, normal requests will be logged at an info level while the test user gets everything logged for a specific request type.

Summary

Writing services requires a lot of supporting code and networking that monolithic applications don't need. And the deployment and management of services are also more involved. So why would anybody design a solution that uses services instead of putting everything into a single monolithic application? Because services can help simplify the design of your applications, especially for very large applications. And because services run on distributed computers, you can scale a solution and increase reliability. Releasing changes and new features in your solution also becomes easier with services because each service can be tested and updated by itself. You don't have to test one giant application and release everything all at once.

This chapter explored some of the different testing challenges with services and what can be tested. You were introduced to a simple service that skips routing and networking and goes straight to the core of what makes a service: the ability to start the service and handle requests.

The simple service developed in this chapter ties together the testing and the logging libraries, which are both used in the service. You can follow a similar project structure when designing your own projects that need to use both libraries.

The next chapter will explore the difficulties of using multiple threads in your testing. We'll test the logging library to make sure it's thread safe, learn what thread safety means, and explore how to test a service with multiple threads.

15

How to Test With Multiple Threads

Multi-threading is one of the most difficult aspects of writing software. Something that's often overlooked is how we can test multiple threads. And can we use TDD to help design software that uses multiple threads? Yes, TDD can help and you'll find useful and practical guidance in this chapter that will show you how to use TDD with multiple threads.

The main topics in this chapter are as follows:

- Using multiple threads in tests
- Making the logging library thread-safe
- The need to justify multiple threads
- Changing the service return type
- Making multiple service calls
- How to test multiple threads without sleep
- Fixing one last problem detected with logging

First, we'll examine what problems you'll find when using multiple threads in your tests. You'll learn how to use a special helper class in the testing library to simplify the extra steps needed when testing with multiple threads.

Once we can use multiple threads inside of a test, we'll use that ability to call into the logging library from multiple threads at the same time and see what happens. I'll give you a hint: some changes will need to be made to the logging library to make the library behave well when called from multiple threads.

Then, we'll go back to the simple service we developed in the previous chapter and you'll learn how to use TDD to design a service that uses multiple threads in a way that can support reliable testing.

We'll be working with each project in turn in this chapter. First, we will be using the testing library project. Then, we'll switch over to the logging library project. Finally, we'll use the simple service project.

Technical requirements

All the code in this chapter uses standard C++, which builds on any modern C++ 20 or later compiler and standard library. The code in this chapter uses all three projects developed in this book: the testing library from *Part 1, Testing MVP*, the logging library from *Part 2, Logging Library*, and the simple service from the previous chapter.

You can find all the code for this chapter in this book's GitHub repository: `https://github.com/PacktPublishing/Test-Driven-Development-with-CPP`.

Using multiple threads in tests

Adding multiple threads to your tests presents challenges that you need to be aware of. I'm not talking about running the tests themselves in multiple threads. The testing library registers and runs the tests and it will remain single-threaded. What you need to understand are the problems that can arise when multiple threads are created inside of a test.

To understand these problems, let's create a test that uses multiple threads so that you can see exactly what happens. We'll be working with the unit test library project in this section so, first, add a new test file called `Thread.cpp`. The project structure should look like this after you've added the new file:

```
MereTDD project root folder
    Test.h
    tests folder
        main.cpp
        Confirm.cpp
        Creation.cpp
        Hamcrest.cpp
        Setup.cpp
        Thread.cpp
```

Inside the `Thread.cpp` file, add the following code:

```
#include "../Test.h"

#include <atomic>
#include <thread>

using namespace MereTDD;

TEST("Test can use additional threads")
```

```
{
    std::atomic<int> count {0};
    std::thread t1([&count]()
    {
        for (int i = 0; i < 100'000; ++i)
        {
            ++count;
        }
        CONFIRM_THAT(count, NotEquals(100'001));
    });
    std::thread t2([&count]()
    {
        for (int i = 0; i < 100'000; ++i)
        {
            --count;
        }
        CONFIRM_THAT(count, NotEquals(-100'001));
    });

    t1.join();
    t2.join();
    CONFIRM_THAT(count, Equals(0));
}
```

The preceding code includes atomic so that we can safely modify a count variable from multiple threads. We need to include thread to bring in the definition of the thread class. The test creates two threads. The first thread increments count, while the second thread decrements the same count. The final result should return count to zero because we increment and decrement the same number of times.

If you build and run the test application, everything will pass. The new test causes no problem at all. Let's change the third CONFIRM_THAT macro so that we can try to confirm that count is not equal to 0 at the end of the test, like so:

```
    t1.join();
    t2.join();
    CONFIRM_THAT(count, NotEquals(0));
```

With this change, the test fails with this result:

```
------- Test: Test can use additional threads
Failed confirm on line 30
    Expected: not 0
    Actual   : 0
```

So far, we have a test that uses multiple threads and it works as expected. We added some confirmations that can detect and report when a value does not match the expected value. You might be wondering what problems multiple threads can cause when the threads seem to be working okay so far.

Here's the quick answer: creating one or more threads inside of a test causes no problem at all – that is, assuming that the threads are managed correctly such as making sure they are joined before the test ends. Confirmations work as expected from the main test thread itself. You can even have confirmations inside the additional threads. One type of problem comes when a confirmation inside one of the additional threads fails. To see this, let's put the final confirmation back to Equals and change the first confirmation to Equals too, like so:

```
for (int i = 0; i < 100'000; ++i)
{
    ++count;
}
CONFIRM_THAT(count, Equals(100'001));
```

count should never reach 100'001 because we only increment 100'000 times. The confirmation always passed before this change, which is why it did not cause a problem. But with this change, the confirmation will fail right away. If this was a confirmation in the main test thread, then the failure would cause the test to fail with a summary message that describes the problem. But we're not in the main test thread now.

Remember that failed confirmations throw exceptions and that an unhandled exception inside of a thread will terminate an application. When we confirm that the count equals 100'001, we cause an exception to be thrown. The main test thread is managed by the testing library and the main thread is ready to catch any confirmation exceptions so that they can be reported. However, our additional thread inside the test lambda has no protection against thrown exceptions. So, when we build and run the test application, it terminates like this:

```
------- Test: Test can use additional threads
terminate called after throwing an instance of
'MereTDD::ActualConfirmException'
Abort trap: 6
```

You might get a slightly different message, depending on what computer you're using. What you won't get is a test application that runs and reports the results of all the tests. The application terminates soon after the confirmation inside the additional thread fails and throws an exception.

Other than confirmations inside a thread failing and throwing exceptions, are there any other problems with using multiple threads inside of a test? Yes. Threads need to be managed properly – that is, we need to make sure they are either joined or detached before going out of scope. You're unlikely to need to detach a thread that was created in a test, so you're left with making sure that all the threads created inside of a test are joined before the test ends. Notice that the test we're using manually joins both threads.

If the test has other confirmations, then you need to be sure that a failed confirmation doesn't cause the test to skip the thread joins. This is because leaving a test without joining will also cause the application to terminate. Let's see this by putting the first confirmation back to using NotEquals so that it will not cause any problems. Then, we will add a new confirmation that will fail before the joins:

```
CONFIRM_TRUE(false);
t1.join();
t2.join();
CONFIRM_THAT(count, Equals(0));
```

The confirmations inside the additional threads no longer cause any problems. However, the new CONFIRM_TRUE confirmation will cause the joins to be skipped. The result is another termination:

```
------- Test: Test can use additional threads
terminate called without an active exception
Abort trap: 6
```

We're not going to do anything to help solve this second type of termination. You'll need to make sure that any threads that are created are joined properly. You might want to use the new *jthread* in C++20, which will make sure that the threads are joined. Alternatively, you might just need to be careful about where you put confirmations in the main test thread to make sure that all the joins happen first.

We can remove the CONFIRM_TRUE confirmation now so that we can focus on fixing the first problem of confirmations failing inside the threads.

What can we do to fix this problem? We could put a try/catch block in the thread, which would at least stop the termination:

```
TEST("Test can use additional threads")
{
    std::atomic<int> count {0};
    std::thread t([&count]()
```

```
        {
            try
            {
                for (int i = 0; i < 100'000; ++i)
                {
                    ++count;
                }
                CONFIRM_THAT(count, NotEquals(100'001));
            }
            catch (...)
            { }
        });

        t.join();
        CONFIRM_THAT(count, Equals(100'000));
    }
```

To simplify the code, I removed the second thread. The test now uses a single additional thread to increment the count. The result after the thread finishes is that count should be equal to 100'000. At no point should count reach 100'001, which is confirmed inside the thread. Let's say we change the confirmation inside the thread so that it will fail:

```
        CONFIRM_THAT(count, Equals(100'001));
```

Here, the exception is caught and the test fails normally and reports the result. Or does it? Building and running this code shows that all the tests pass. The confirmation inside the thread is detecting the mismatched values but the exception has no way to be reported back to the main test thread. We can't throw anything inside the catch block because that will just terminate the application again.

We know that we can avoid the test application terminating by catching the confirmation exception. And we also know from the first threading test that a confirmation that doesn't throw is also okay. The bigger problem we need to solve is how to let the main test thread know about any confirmation failures in the additional threads that have been created. Maybe we can inform the main thread in the catch block by using a variable passed to the thread.

I want to emphasize this point. If you're creating threads inside of a test simply to divide the work and speed up a test and don't need to confirm anything inside the threads, then you don't need to do anything special. All you need to manage is the normal thread concerns, such as making sure you join all threads before the test ends and that none of the threads have unhandled exceptions. The only reason to use the following guidance is when you want to put confirmations inside of the additional threads.

After trying out a few alternatives, here is what I came up with:

```
TEST("Test can use additional threads")
{
    ThreadConfirmException threadEx;
    std::atomic<int> count {0};
    std::thread t([&threadEx, &count]()
    {
        try
        {
            for (int i = 0; i < 100'000; ++i)
            {
                ++count;
            }
            CONFIRM_THAT(count, Equals(100'001));
        }
        catch (ConfirmException const & ex)
        {
            threadEx.setFailure(ex.line(), ex.reason());
        }
    });

    t.join();
    threadEx.checkFailure();
    CONFIRM_THAT(count, Equals(100'000));
}
```

This is the TDD style. Modify the test until you're happy with the code and then get it working. The test assumes a new exception type called ThreadConfirmException and it creates a local instance called threadEx. The threadEx variable is captured by reference in the thread lambda so that the thread can access threadEx.

The thread can use all the normal confirmations it wants, so long as everything is inside a try block with a catch block that is looking for the ConfirmException type. If a confirmation fails, then it will throw an exception that will be caught. We can use the line number and reason to set a failure mode in the threadEx variable.

Once the thread has finished and we're back in the main thread, we can call another method to check for a failure in the threadEx variable. If a failure was set, then the checkFailure method should throw an exception, just like how a regular confirmation throws an exception.

Because we're back in the main test thread, any confirmation exception that gets thrown will be detected and reported in the test summary report.

Now, we need to implement the `ThreadConfirmException` class in `Test.h`, which can go right after the `ConfirmException` base class, like this:

```
class ThreadConfirmException : public ConfirmException
{
public:
    ThreadConfirmException ()
    : ConfirmException(0)
    { }

    void setFailure (int line, std::string_view reason)
    {
        mLine = line;
        mReason = reason;
    }

    void checkFailure () const
    {
        if (mLine != 0)
        {
            throw *this;
        }
    }
};
```

If we build and run now, then the confirmation inside the thread will detect that `count` does not equal `100'001` and the failure will be reported in the summary results, like this:

```
------- Test: Test can use additional threads
Failed confirm on line 20
    Expected: 100001
    Actual  : 100000
```

The question now is, is there any way to simplify the test? The current test looks like this:

```
TEST("Test can use additional threads")
{
```

```
        ThreadConfirmException threadEx;
        std::atomic<int> count {0};
        std::thread t([&threadEx, &count]()
        {
            try
            {
                for (int i = 0; i < 100'000; ++i)
                {
                    ++count;
                }
                CONFIRM_THAT(count, Equals(100'001));
            }
            catch (ConfirmException const & ex)
            {
                threadEx.setFailure(ex.line(), ex.reason());
            }
        });

        t.join();
        threadEx.checkFailure();
        CONFIRM_THAT(count, Equals(100'000));
}
```

Here, we have a new ThreadConfirmException type, which is good. However, the test author still needs to pass an instance of this type to the thread function, similar to how threadEx is captured by the lambda. The thread function still needs a try/catch block and needs to call setFailure if an exception is caught. Finally, the test needs to check for a failure once it's back in the main test thread. All of these steps are shown in the test.

We might be able to use a few macros to hide the try/catch block, but this seems fragile. The test author will likely have slightly different needs. For example, let's go back to two threads and see what the test will look like with multiple threads. Change the test so that it looks like this:

```
TEST("Test can use additional threads")
{
    std::vector<ThreadConfirmException> threadExs(2);
    std::atomic<int> count {0};
    std::vector<std::thread> threads;
    for (int c = 0; c < 2; ++c)
```

```
    {
        threads.emplace_back(
            [&threadEx = threadExs[c], &count]()
            {
                try
                {
                    for (int i = 0; i < 100'000; ++i)
                    {
                        ++count;
                    }
                    CONFIRM_THAT(count, Equals(200'001));
                }
                catch (ConfirmException const & ex)
                {
                    threadEx.setFailure(ex.line(), ex.reason());
                }
            });
    }

    for (auto & t : threads)
    {
        t.join();
    }
    for (auto const & ex: threadExs)
    {
        ex.checkFailure();
    }
    CONFIRM_THAT(count, Equals(200'000));
}
```

This test is different than the original two-thread test at the beginning of this section. I wrote the test differently to show that there are lots of ways to write a multi-threaded test. Because we have more code inside the thread to handle the confirmation exceptions, I made each thread similar. Instead of one thread incrementing the count while another thread decrements, both threads now increment. Also, instead of naming each thread t1 and t2, the new test puts the threads in a vector. We also have a vector of ThreadConfirmException with each thread getting a reference to its own ThreadConfirmException.

One thing to notice about this solution is that while each thread will fail its confirmation and both `ThreadConfirmationException` instances will have a failure set, only one failure will be reported. In the loop at the end of the test that goes through the `threadExs` collection, the moment one `ThreadConfirmationException` fails the check, an exception will be thrown. I thought about extending the testing library to support multiple failures but decided against the added complexity.

If you have a test with multiple threads, then they will likely be working with different sets of data. If there happens to be an error that causes multiple threads to fail in the same test run, then only one failure will be reported in the test application. Fixing that failure and running again may then report the next failure. It's a little tedious to fix problems one after another but not a likely scenario that justifies the added complexity to the testing library.

The new test structure with two threads highlights the difficulty of creating reasonable macros that can hide all the thread confirmation handling. So far, all three versions of the test have been different. There doesn't seem to be a common way to write multi-threaded tests that we would be able to wrap up in some macros. I think we'll stick with what we have now – a `ThreadConfirmException` type that can be passed to a thread. The thread will need to catch the `ConfirmException` type and call `setFailure`. The main test thread can then check each `ThreadConfirmException`, which will throw if the failure was set. Before we move on, let's change the confirmation inside the thread lambda so that it tests for a count not equal to `200'001`, like this:

```
CONFIRM_THAT(count, NotEquals(200'001));
```

The `NotEquals` confirmation will let the test pass again.

With the understanding you've gained from this section, you'll be able to write tests that use multiple threads inside the test. You can continue to use the same `CONFIRM` and `CONFIRM_THAT` macros to verify the results. The next section will use multiple threads to log messages so that we can make sure that the logging library is thread-safe. You'll also learn what it means for code to be thread-safe.

Making the logging library thread-safe

We don't know if a project that uses the logging library will be trying to log from multiple threads or a single thread. With an application, we're in full control and can choose to use multiple threads or not. But a library, especially a logging library, often needs to be *thread-safe*. This means that the logging library needs to behave well when an application uses the library from multiple threads. Making code thread-safe adds some extra overhead to the code and is not needed if the library will only be used from a single thread.

What we need is a test that calls `log` from multiple threads that are all running at the same time. Let's write a test with the code we have now and see what happens. We're going to be using the logging project in this section and adding a new file to the `tests` folder called `Thread.cpp`. The project structure will look like this with the new file added:

```
MereMemo project root folder
```

```
MereTDD folder
    Test.h
MereMemo folder
    Log.h
    tests folder
        main.cpp
        Construction.cpp
        LogTags.h
        Tags.cpp
        Thread.cpp
        Util.cpp
        Util.h
```

Inside the Thread.cpp file, let's add a test that calls log from several threads, like so:

```cpp
#include "../Log.h"

#include "Util.h"

#include <MereTDD/Test.h>
#include <thread>

TEST("log can be called from multiple threads")
{
    // We'll have 3 threads with 50 messages each.
    std::vector<std::string> messages;
    for (int i = 0; i < 150; ++i)
    {
        std::string message = std::to_string(i);
        message += " thread-safe message ";
        message += Util::randomString();
        messages.push_back(message);
    }

    std::vector<std::thread> threads;
    for (int c = 0; c < 3; ++c)
    {
```

```
            threads.emplace_back(
                [c, &messages]()
                {
                    int indexStart = c * 50;
                    for (int i = 0; i < 50; ++i)
                    {
                        MereMemo::log() << messages[indexStart + i];
                    }
                });
        }

        for (auto & t : threads)
        {
            t.join();
        }
        for (auto const & message: messages)
        {
            bool result = Util::isTextInFile(message,
                "application.log");
            CONFIRM_TRUE(result);

        }
    }
```

This test does three things. First, it creates 150 messages. We'll get the messages ready before we start the threads so that the threads will be able to call log as quickly as possible many times in a loop.

Once the messages are ready, the test starts 3 threads, and each thread will log part of the messages that have already been formatted. The first thread will log messages 0 to 49. The second thread will log messages 50 to 99. Finally, the third thread will log messages 100 to 149. We don't do any confirmations in the threads.

Once everything has been logged and the threads have been joined, then the test confirms that all 150 messages appear in the log file.

Building and running this will almost certainly fail. This type of test goes against one of the points that makes a good test, as explained in *Chapter 8, What Makes A Good Test?* The reason this is not the best type of test is that the test is not completely reproducible. Each time the test application is run, you'll get a slightly different result. You might even find that this test causes other tests to fail!

Even though we're not basing the behavior of the test on random numbers, we're using threads. And thread scheduling is unpredictable. The only way to make this test mostly reliable is to log many messages like we're doing already. The test does everything it can to set the threads up for conflicts. This is why the messages are preformatted. I wanted the threads to immediately go into a loop of logging messages and not spend any extra time formatting messages.

When the test fails, it's because the log file is jumbled. One portion of the log file looks like this for one of my test runs:

```
2022-08-16T04:54:54.635 100 thread-safe message 4049
2022-08-16T04:54:54.635 100 thread-safe message 4049
2022-08-16T04:54:54.635 0 thread-safe message 8866
2022-08-16T04:54:54.637 101 thread-safe message 8271
2022-08-16T04:54:54.637 1 thread-safe message 3205
2022-08-16T04:54:54.637 102 thread-safe message 7514
2022-08-16T04:54:54.637 51 thread-safe message 7405
2022-08-16T04:54:54.637 2 thread-safe message 5723
2022-08-16T04:54:54.637 52 thread-safe message 4468
2022-08-16T04:54:54.637 52 thread-safe message 4468
```

I removed the `color` and `log_level` tags so that you can see the messages better. The first thing you'll notice is that some messages are repeated. Number `100` appears twice, and number `50` seems to be missing completely.

To be honest, I expected the log file to be even more jumbled than it is. The interleaving between message groups `0-49` and `50-99` and `100-149` is to be expected. We do have three threads running at the same time. For example, once message number `51` is logged, we should expect to have already seen number `50`.

Let's fix the logging code to get the test to pass. It still won't be the best test but it will have a good chance of finding a bug if the logging library is not thread-safe.

The fix is simple: we need a mutex and then we need to lock the mutex. First, let's include the `mutex` standard header at the top of `Log.h`, like this:

```
#include <algorithm>
#include <chrono>
#include <ctime>
#include <filesystem>
#include <fstream>
#include <iomanip>
#include <map>
#include <memory>
#include <mutex>
```

```
#include <ostream>
#include <sstream>
#include <string>
#include <string_view>
#include <vector>
```

Then, we need a place to place a global mutex. Since the logging library is a single header file, we can't declare a global variable without getting a linker error. We might be able to declare a global mutex as inline. This is a new feature in C++ that I haven't used that lets you declare inline variables, just like how we can declare inline functions. I'm more comfortable with a function that uses a static variable. Add the following function to the top of Log.h, right after the opening namespace of MereMemo:

```
inline std::mutex & getLoggingMutex ()
{
    static std::mutex m;
    return m;
}
```

Now, we need to lock the mutex at the proper spot. At first, I added a lock to the log function, but that had no effect. This is because the log function returns a LogStream without actually doing any logging. So, the log function obtained the lock and then released the lock before any logging happened. The logging is done in the LogStream destructor, so that's where we need to put the lock:

```
~LogStream ()
{
    if (not mProceed)
    {
        return;
    }

    const std::lock_guard<std::mutex>
        lock (getLoggingMutex ());

    auto & outputs = getOutputs ();
    for (auto const & output: outputs)
    {
        output->sendLine (this->str ());
    }
}
```

The lock tries to obtain the mutex and will block if another thread already owns the mutex. Only one thread at a time can proceed after the lock and the lock is released after the text is sent to all the outputs.

If we build and run, the threading problem will be fixed. However, when I ran the test application, one of the tests failed. At first, I thought there was still a problem with the threads, but the failure was in another test. This is the test that failed:

```
TEST("Overridden default tag not used to filter messages")
{
    MereTDD::SetupAndTeardown<TempFilterClause> filter;
    MereMemo::addFilterLiteral(filter.id(), info);

    std::string message = "message ";
    message += Util::randomString();
    MereMemo::log(debug) << message;

    bool result = Util::isTextInFile(message,
        "application.log");
    CONFIRM_FALSE(result);
}
```

This test has nothing to do with the multiple thread test. So, why did it fail? Well, the problem is that this test is confirming that a particular message does not appear in the log file. But the message is just the word `"message "`, followed by a random number string. We just added an extra 150 logged messages, which all have the same text followed by a random number string.

We have a problem with the tests themselves. The tests can sometimes fail due to random numbers. The problem wasn't noticed when we had a few log messages but it's more noticeable now that we have many more chances for duplicate random numbers.

We could either increase the size of the random number strings added to each log message or make the tests more specific so that they all use a different base message string.

At this point, you might be wondering why my test has a simple base message when we've been using unique messages in each test ever since the logging library was first created in *Chapter 9, Using Tests*. That's because the code starting in *Chapter 9, Using Tests*, originally did have simple, common log messages. I could have left those common messages as-is and waited until now to have you go back and change all of them. However, I edited the chapters to fix the problem from the beginning. It seems like a waste to go through all the tests now just to change a string. Therefore, I added an explanation to *Chapter 9, Using Tests*. We don't need to change any of the test messages now because they've already been fixed.

Okay, back to the threading – the new test passes now and the sample from the log file looks much better:

```
2022-08-16T06:20:36.807 0 thread-safe message 6269
2022-08-16T06:20:36.807 50 thread-safe message 1809
2022-08-16T06:20:36.807 100 thread-safe message 6297
2022-08-16T06:20:36.808 1 thread-safe message 848
2022-08-16T06:20:36.808 51 thread-safe message 4103
2022-08-16T06:20:36.808 101 thread-safe message 5570
2022-08-16T06:20:36.808 2 thread-safe message 6156
2022-08-16T06:20:36.809 102 thread-safe message 4213
2022-08-16T06:20:36.809 3 thread-safe message 6646
```

Again, this sample has been modified to remove the `color` and `log_level` tags. This change makes each line shorter so that you can see the messages better. The messages within each thread are ordered, even though the messages are mixed between threads – that is, message number 0 is followed at some point by message number 1 and then by number 2; message number 50 is followed later by number 51, and message number 100 is followed by number 101. Each following numbered message might not immediately follow the previous message. This sample looks better because there are no duplicates and no missing messages.

One final thought is about the thread-safety of the logging library. We tested that multiple threads can all safely call `log` without worrying about problems. But we didn't test if multiple threads can manage default tags or filtering, or add new outputs. The logging library will likely need more work to be fully thread-safe. It will work for our purposes for now.

Now that the logging library is mostly thread-safe, the next section will go back to the `SimpleService` project and begin exploring how to test code that uses multiple threads.

The need to justify multiple threads

So far in this chapter, you've learned how to write tests that use multiple threads and how to use these extra threads to test the logging library. The logging library doesn't use multiple threads itself, but we needed to make sure that the logging library is safe to use with multiple threads.

The remainder of this chapter will provide some guidance on how to test code that does use multiple threads. To test multi-threaded code, we need some code that uses multiple threads. For this, we'll use the `SimpleService` project from the previous chapter.

We need to modify the simple service so that it uses multiple threads. Right now, the simple service is an example of a greeting service that responds to a greeting request with a reply based on the user making the request being identified. There's not much of a need for multiple threads in a greeting service. We're going to need something different.

This brings us to the first guidance: we need to make sure there is a valid need for multiple threads before we try to add multiple threads. Writing multi-threaded code is hard and should be avoided if only a single thread is needed. If you only need a single thread, then make sure that you follow the advice from the previous section and make your code thread-safe if it will be used by multiple threads.

What you want to do is write as much of your code as possible so that it's single-threaded. If you can identify a particular way to calculate a result that only needs some input data to arrive at an output, then make that a single-threaded calculation if possible. If the amount of input data is large and can be divided and calculated separately, then break up the input and pass smaller pieces to your calculation. Keep the calculation single-threaded and focused on working with the input provided. Then, you can create multiple threads where each thread is given a portion of the input data to calculate. This will separate your multi-threaded code from your calculations.

Isolating your single-threaded code will let you design and test the code without you having to worry about thread management. Sure, you might need to make sure the code is thread-safe, but that's easier when thread-safety is all you need to worry about.

Testing multiple threads is harder because of the randomness of the thread scheduling. If possible, try to avoid clunky methods such as *sleeping* to coordinate tests. You want to avoid putting actual code threads to sleep to coordinate the order between threads. When a thread goes to sleep, it stops running for a while, depending on how long of a delay is specified in the sleep call. Other threads that are not sleeping can then be scheduled to run.

We'll design the code in this chapter to let the test control the thread's synchronization so that we can remove the randomness and make the tests predictable. Instead of starting this section with a test, let's look at a modified service that has a reason to use multiple threads. The modified `handleRequest` method looks like this:

```cpp
std::string SimpleService::Service::handleRequest (
    std::string const & user,
    std::string const & path,
    std::string const & request)
{

    MereMemo::log(debug, User(user), LogPath(path))
        << "Received: " << Request(request);

    std::string response;
    if (request == "Calculate")
    {
        response = "token";
    }
    if (request == "Status")
```

```
    {
        response = "result";
    }
    else
    {
        response = "Unrecognized request.";
    }

    MereMemo::log(debug, User(user), LogPath(path))
        << "Sending: " << Response(response);
    return response;
}
```

When following TDD, you'll normally want to start with tests first. So, why am I showing you a modified service first? Because our goal is to test multi-threaded code. In your projects, you should avoid the desire to use some technology without having a good reason. Our reason is that we need an example to learn from. So, we're starting with a backward need to use multi-threading.

I tried to think of a good reason for a greeting service to use multiple threads and nothing came to mind. So, we're going to change the service to something a little more complicated; I want to explain this new idea before we begin writing tests.

The new service is still as simple as I can make it. We'll continue ignoring all the networking and message routing. We'll need to change the request and response types to structs and we'll continue to ignore serializing the data structs for transmission to and from the service.

The new service will simulate the calculation of a difficult problem. One valid reason to create a new thread is to let the new thread perform some work while the original thread continues what it was doing. The idea of the new service is that a `Calculate` request can take a long time to complete and we don't want the caller to time out while waiting for the result. So, the service will create a new thread to perform the calculation and immediately return a token to the caller. The caller can use this token to call back into the service with a different `Status` request, which will check on the progress of the calculation that was just begun. If the calculation is not done yet, then the response to the `Status` request will let the caller know approximately how much has been completed. If the calculation is done, then the response will contain the answer.

We now have a justification for multiple threads and can write some tests. Let's take care of an unrelated test that should have been added already. We want to make sure that anybody calling the service with an unrecognized request will get an unrecognized response. Put the following test in the `Message.cpp` file in the `tests` folder of the `SimpleService` project:

```
TEST_SUITE("Unrecognized request is handled properly", "Service
1")
```

```
{
    std::string user = "123";
    std::string path = "";
    std::string request = "Hello";
    std::string expectedResponse = "Unrecognized request.";

    std::string response = gService1.service().handleRequest(
        user, path, request);
    CONFIRM_THAT(response, Equals(expectedResponse));
}
```

I put this test at the top of Message.cpp. All it does is send the previous greeting request but with an unrecognized expected response.

Let's also change the name of the test suite to "Calculation Service" like this in SetupTeardown.cpp:

```
MereTDD::TestSuiteSetupAndTeardown<ServiceSetup>
gService1("Calculation Service", "Service 1");
```

Now, let's remove the greeting test and add the following simple test, which makes sure we get something other than the unrecognized response:

```
TEST_SUITE("Calculate request can be sent and recognized",
"Service 1")
{
    std::string user = "123";
    std::string path = "";
    std::string request = "Calculate";
    std::string unexpectedResponse = "Unrecognized request.";

    std::string response = gService1.service().handleRequest(
        user, path, request);
    CONFIRM_THAT(response, NotEquals(unexpectedResponse));
}
```

This test is the opposite of the unrecognized test and makes sure that the response is something other than unrecognized. Normally, it's better to confirm that a result matches what you expect to happen instead of confirming that a result is not what you don't expect. A double negative is not only harder to think about, but can lead to problems because it's not possible to catch all the ways something can go wrong. By confirming what you want to happen, you can eliminate all the possible error conditions, which are too many to catch individually.

This test is a little different, though. We're not interested in the response. The test only intends to confirm that the request was recognized. Confirming that the response is not unrecognized is appropriate, even though it seems similar to the double negative trap we just described.

Building and running this code shows that the unrecognized test passes but the `Calculate` request fails:

```
Running 1 test suites
--------------- Suite: Service 1
------- Setup: Calculation Service
Passed
------- Test: Unrecognized request is handled properly
Passed
------- Test: Calculate request can be sent and recognized
Failed confirm on line 30
    Expected: not Unrecognized request.
    Actual  : Unrecognized request.
------- Teardown: Calculation Service
Passed
-----------------------------------
Tests passed: 3
Tests failed: 1
```

It seems that we're getting an unrecognized response for a request that should be valid. This is the value of adding simple tests at the beginning of a project. The tests help catch simple errors right away. The problem is in the `handleRequest` method. I added the second check for a valid request by copying the first check and forgot to change the `if` statement to an `else if` statement. The fix for this is as follows:

```
if (request == "Calculate")
{
    response = "token";
}
else if (request == "Status")
{
    response = "result";
}
```

```
    else
    {
        response = "Unrecognized request.";
    }
```

To continue further, we're going to send and receive more than strings. When we send a `Calculate` request, we should get back a token value that we can pass to the `Status` request. The `Status` response should then contain either the answer or an estimate of how much progress has been made. Let's take this one step at a time and define the `Calculate` request and response structures. Add the following two struct definitions to the top of `Service.h` inside the `SimpleService` namespace:

```
struct CalculateRequest
{
    int mSeed;
};

struct CalculateResponse
{
    std::string mToken;
};
```

This will let us pass some initial value to be calculated; in return, we will get a token that we can use to eventually get the answer. But we have a problem. If the `Calculate` request is changed to return a struct, then that will break the existing test, which expects a string. We should change the tests so that they use the structs, but that leads to another problem: most of the time, we need to return the correct response struct. And we need to return an error response for error cases.

What we need is a response that can represent both a good response and an error response. Since we're going to have a response that can serve multiple purposes, why not let it also handle a struct for the `Status` response? This means we'll have a single response type that can be either an error response, a calculate response, or a status response. And since we have a multi-purpose response type, why not create a multi-purpose request type? Let's change the tests.

We're going to use `std::variant` to hold the different types of requests and responses. We can remove the test that sent a request string that was not valid. We can still get an invalid request but only with mismatched service versions between the caller and the service. That's a little more involved, so we'll ignore the possibility that a service can be called with a different idea of what requests are available than the service knows about. If you're writing a real service, then this is a possibility that needs to be addressed and tested. You'll probably want to use something different than a variant too. A good choice would be something such as Google's *Protocol Buffers*, where the service would accept Protocol Buffer messages. While using Protocol Buffers is a better choice than simple structs, the design is also a lot more complicated and would make this explanation much longer.

We'll have a single test in `Message.cpp` that will look like this:

```
TEST_SUITE("Calculate request can be sent", "Service 1")
{
    std::string user = "123";
    std::string path = "";

    SimpleService::RequestVar request =
        SimpleService::CalculateRequest {
            .mSeed = 5
        };
    std::string emptyResponse = "";
    std::string response = gService1.service().handleRequest(
        user, path, request);
    CONFIRM_THAT(response, NotEquals(emptyResponse));
}
```

This test focuses on the request type first and leaves the response type as a string. We'll make the changes one step at a time. This is especially good advice when working with `std::variant` because it can be challenging if you're not familiar with variants. We'll have a variant type called `RequestVar` that can be initialized with a specific request type. We're initializing the request with a `CalculateRequest` and using the *designated initializer* syntax to set the mSeed value. The designated initializer syntax is fairly new to C++. It lets us set data member values based on the name by putting a dot in front of the data member's name.

Now, let's define the request types in `Service.h`:

```
#ifndef SIMPLESERVICE_SERVICE_H
#define SIMPLESERVICE_SERVICE_H

#include <string>
#include <variant>

namespace SimpleService
{

struct CalculateRequest
{
    int mSeed;
```

```
};

struct StatusRequest
{
    std::string mToken;
};

using RequestVar = std::variant<
    CalculateRequest,
    StatusRequest
    >;
```

Note that we need to include the standard variant header file. The RequestVar type can now only be either a CalculateRequest or a StatusRequest. We need to make one more change in Service.h to the handleRequest method in the Service class:

```
class Service
{
public:
    void start ();

    std::string handleRequest (std::string const & user,
        std::string const & path,
        RequestVar const & request);
};
```

The Service.cpp file needs to be changed so that it updates the handleRequest method, like this:

```
std::string SimpleService::Service::handleRequest (
    std::string const & user,
    std::string const & path,
    RequestVar const & request)
{
    std::string response;
    if (auto const * req = std::get_
       if<CalculateRequest>(&request))
    {
        MereMemo::log(debug, User(user), LogPath(path))
```

```
            << "Received Calculate request for: "
            << std::to_string(req->mSeed);
        response = "token";
    }
    else if (auto const * req = std::get_
            if<StatusRequest>(&request))
    {

        MereMemo::log(debug, User(user), LogPath(path))
            << "Received Status request for: "
            << req->mToken;
        response = "result";

    }
    else
    {

        response = "Unrecognized request.";

    }


    MereMemo::log(debug, User(user), LogPath(path))
        << "Sending: " << Response(response);
    return response;

}
```

The updated `handleRequest` method continues to check for an unknown request type. All the responses are strings that will need to change. We're not looking at the seed or token values yet, but we have enough that can be built and tested.

Now that the single test passes, in the next section, we will look at the responses and use structs instead of response strings.

Changing the service return type

We'll be making a similar change in this section to move away from strings and use a struct in the service request handling. The previous section changed the service request type; this section will change the service return type. We need to make these changes so that we can get the service to a level of functionality where it can support the need for an additional thread.

The `SimpleService` project that we're using started as a greeting service and I could not think of any reason for such a simple service to need another thread. We started adapting the service to a calculation service in the previous section; now, we need to modify the return types that the service returns when handling requests.

First, let's define the return type structs in `Service.h`, which come right after the request types. Add the following code to `Service.h`:

```cpp
struct ErrorResponse
{
    std::string mReason;
};

struct CalculateResponse
{
    std::string mToken;
};

struct StatusResponse
{
    bool mComplete;
    int mProgress;
    int mResult;
};

using ResponseVar = std::variant<
    ErrorResponse,
    CalculateResponse,
    StatusResponse
    >;
```

These structs and the variant are following the same pattern that was used for the requests. One small difference is that we now have an `ErrorResponse` type, which will be returned for any errors. We can modify the test in `Message.cpp` so that it looks like this:

```cpp
TEST_SUITE("Calculate request can be sent", "Service 1")
{
    std::string user = "123";
    std::string path = "";

    SimpleService::RequestVar request =
        SimpleService::CalculateRequest {
            .mSeed = 5
```

```
        };
    auto const responseVar = gService1.service().handleRequest(
        user, path, request);
    auto const response =
        std::get_
  if<SimpleService::CalculateResponse>(&responseVar);
    CONFIRM_TRUE(response != nullptr);
}
```

This test will call the service as it did previously with a calculate request; the response that comes back is tested to see if it is a calculate response.

For the code to compile, we need to change the handleRequest declaration in Service.h so that it returns the new type, like this:

```
class Service
{
public:
    void start ();

    ResponseVar handleRequest (std::string const & user,
        std::string const & path,
        RequestVar const & request);
};
```

Then, we need to change the implementation of handleRequest in Service.cpp:

```
SimpleService::ResponseVar
SimpleService::Service::handleRequest (
    std::string const & user,
    std::string const & path,
    RequestVar const & request)
{
    ResponseVar response;
    if (auto const * req = std::get_
        if<CalculateRequest>(&request))
    {
        MereMemo::log(debug, User(user), LogPath(path))
            << "Received Calculate request for: "
            << std::to_string(req->mSeed);
```

```
        response = SimpleService::CalculateResponse {
            .mToken = "token"
        };
    }
    else if (auto const * req = std::get_
            if<StatusRequest>(&request))
    {
        MereMemo::log(debug, User(user), LogPath(path))
            << "Received Status request for: "
            << req->mToken;
        response = SimpleService::StatusResponse {
            .mComplete = false,
            .mProgress = 25,
            .mResult = 0
        };
    }
    else
    {
        response = SimpleService::ErrorResponse {
            .mReason = "Unrecognized request."
        };
    }

    return response;
}
```

The code is getting a little more complicated. I removed the log at the end, which was used to log the response before returning. We could put the log back in but that would require the ability to convert a `ResponseVar` into a string. Alternatively, we would need to log the response in multiple places like the code does for the request. That's a detail that we can skip.

The new `handleRequest` method does almost the same things it used to do except that it now initializes a `ResponseVar` type instead of returning a string. This allows us to return different types with more detailed information than before when we were returning a string for both the requests and the error.

To add a test for an unrecognized request, we would need to add a new request type to `RequestVar` but ignore the new request type in the `if` statements inside the `handleRequest` method. We're going to skip that test too because we really should be using something other than a `std::variant`.

The only reason we're using `std::variant` for this example is to avoid extra complexity. We're trying to get the code ready to support another thread.

In the next section, we will add a test that uses both request types. The first request will begin a calculation, while the second request will check the status of the calculation and get the result when the calculation is complete.

Making multiple service calls

If you're considering using multiple threads to speed up a calculation, then I recommend that you get the code tested and working with a single thread before taking on the additional complexity of multiple threads.

For the service we're working on, the reason to add a second thread is not to increase the speed of anything. We need to avoid a timeout for a calculation that might take a long time. The additional thread we're going to add is not designed to make the calculation any faster. Once we get the calculation working with one additional thread, we can consider adding more threads to speed up the calculation.

The need to create a thread to do some work while the original thread continues with something else is common. This is not an optimization that should be done later. This is part of the design and the additional thread should be included from the very beginning.

Let's begin by adding a new test to `Message.cpp` that looks like this:

```cpp
TEST_SUITE("Status request generates result", "Service 1")
{
    std::string user = "123";
    std::string path = "";

    SimpleService::RequestVar calcRequest =
        SimpleService::CalculateRequest {
            .mSeed = 5
        };
    auto responseVar = gService1.service().handleRequest(
        user, path, calcRequest);
    auto const calcResponse =
        std::get_if<SimpleService::CalculateResponse>
        (&responseVar);
    CONFIRM_TRUE(calcResponse != nullptr);

    SimpleService::RequestVar statusRequest =
```

```
            SimpleService::StatusRequest {
                .mToken = calcResponse->mToken
            };
        int result {0};
        for (int i = 0; i < 5; ++i)
        {
            responseVar = gService1.service().handleRequest(
                user, path, statusRequest);
            auto const statusResponse =
                std::get_if<SimpleService::StatusResponse>
                (&responseVar);
            CONFIRM_TRUE(statusResponse != nullptr);

            if (statusResponse->mComplete)
            {
                result = statusResponse->mResult;
                break;
            }
        }
        CONFIRM_THAT(result, Equals(50));
}
```

All the code is already in place for this new test to compile. Now, we can run the tests to see what happens. The test will fail, as follows:

```
Running 1 test suites
-------------- Suite: Service 1
------- Setup: Calculation Service
Passed
------- Test: Calculate request can be sent
Passed
------- Test: Status request generates result
Failed confirm on line 62
    Expected: 50
    Actual  : 0
------- Teardown: Calculation Service
Passed
```

```
--------------------------------------
Tests passed: 3
Tests failed: 1
```

What does the test do? First, it creates a calculate request 1 and gets back a hardcoded token value. There is no calculation for when the service begins yet, so when we make a status request with the token, the service responds with a hardcoded response that says the calculation is not done yet. The test is looking for a status response that says the calculation is complete. The test tries making a status request five times before giving up, which causes the confirmation at the end of the test to fail because we didn't get the expected result. Note that even trying multiple times is not the best way to proceed. Threads are unpredictable and your computer may make all five attempts before the service can complete the request. You might need to increase the number of attempts if your test continues to fail or wait for a reasonable amount of time. Our calculation will eventually multiply the seed by 10. So, when we give an initial seed of 5, we should expect a final result of 50.

We need to implement the calculation and status request handling in the service so that we can use a thread to get the test to pass. The first thing we need to do is include mutex, thread, and vector at the top of Service.cpp. We also need to add an unnamed namespace, like this:

```
#include "Service.h"

#include "LogTags.h"

#include <MereMemo/Log.h>
#include <mutex>
#include <thread>
#include <vector>


namespace
{
}
```

We're going to need some locking so that we don't try to read the calculation status while the status is being updated by a thread. To do the synchronization, we'll use a mutex and a lock, as we did in the logging library. There are other designs you might want to explore, such as locking data for different calculation requests separately. We're going to use a simple approach and have a single lock for everything. Add the following function inside the unnamed namespace:

```
        std::mutex & getCalcMutex ()
        {
            static std::mutex m;
```

```
        return m;
}
```

We need something to keep track of the completion status, the progress, and the result for each calculation request. We'll create a class to hold this information called `CalcRecord` inside the unnamed namespace, right after the `getCalcMutex` function, like this:

```
class CalcRecord
{
public:
    CalcRecord ()
    { }

    CalcRecord (CalcRecord const & src)
    {
        const std::lock_guard<std::mutex>
            lock(getCalcMutex());
        mComplete = src.mComplete;
        mProgress = src.mProgress;
        mResult = src.mResult;
    }

    void getData (bool & complete, int & progress, int &
                result)
    {
        const std::lock_guard<std::mutex>
            lock(getCalcMutex());
        complete = mComplete;
        progress = mProgress;
        result = mResult;
    }

    void setData (bool complete, int progress, int result)
    {
        const std::lock_guard<std::mutex>
            lock(getCalcMutex());
        mComplete = complete;
        mProgress = progress;
```

```
            mResult = result;
        }

        CalcRecord &
        operator = (CalcRecord const & rhs) = delete;

    private:
        bool mComplete {false};
        int mProgress {0};
        int mResult {0};
    };
```

It looks like there's a lot more to this class, but it's fairly simple. The default constructor doesn't need to do anything because the data members already define their default values. The only reason we need a default constructor is that we also have a copy constructor. And the only reason we need a copy constructor is so that we can lock the mutex before copying the data members.

Then, we have a method to get the data members all at once and another method to set the data members. Both the getter and the setter need to acquire the lock before proceeding.

There should be no need to assign one CalcRecord to another, so the assignment operator has been deleted.

The last thing we need in the unnamed namespace is a vector of CalcRecord, like this:

```
    std::vector<CalcRecord> calculations;
```

We're going to add a CalcRecord to the calculations collection every time a calculation request is made. A real service would want to clean up or reuse CalcRecord entries.

We need to modify the request handling in Service.cpp so that a thread gets created to use a new CalcRecord every time we get a calculation request, like this:

```
    if (auto const * req = std::get_
        if<CalculateRequest>(&request))
    {
        MereMemo::log(debug, User(user), LogPath(path))
            << "Received Calculate request for: "
            << std::to_string(req->mSeed);

        calculations.emplace_back();
        int calcIndex = calculations.size() - 1;
```

```
std::thread calcThread([calcIndex] ()
{
    calculations[calcIndex].setData(true, 100, 50);
});
calcThread.detach();
response = SimpleService::CalculateResponse {
    .mToken = std::to_string(calcIndex)
};
}
```

What happens when we get a calculation request? First, we add a new `CalcRecord` to the end of the `calculations` vector. We'll use the index of `CalcRecord` as the token that gets returned in the response. This is the simplest design I could think of to identify a calculation request. A real service would want to use a more secure token. The request handler then starts a thread to do the calculation and detaches from the thread.

Most threading code that you'll write will create a thread and then join the thread. It's not very common to create a thread and then detach from the thread. Alternatively, you can use a pool of threads when you want to do some work and not worry about joining. The reason for detaching is that I wanted the most simple example without bringing in thread pools.

The thread itself is very simple because it immediately sets `CalcRecord` to complete with a progress of `100` and a result of `50`.

We can build and run the test application now, but we will get the same failure we did previously. That's because the status request handling still returns a hardcoded response. We need to modify the request handler like this for the status request:

```
else if (auto const * req = std::get_
        if<StatusRequest>(&request))
{
    MereMemo::log(debug, User(user), LogPath(path))
        << "Received Status request for: "
        << req->mToken;

    int calcIndex = std::stoi(req->mToken);
    bool complete;
    int progress;
    int result;
    calculations[calcIndex].getData(complete, progress,
                        result);
```

```
        response = SimpleService::StatusResponse {
            .mComplete = complete,
            .mProgress = progress,
            .mResult = result
        };
    }
```

With this change, the status request converts the token into an index that it uses to find the correct `CalcRecord`. Then, it gets the current data from `CalcRecord` to be returned in the response.

You may also want to consider adding sleep to the test loop that attempts five service call requests so that the total time given to the service is reasonable. The current test will fail if all five attempts are made quickly before the service has time to complete even a simple calculation.

All the tests pass after building and running the test application. Are we done now? Not yet. All of these changes let the service calculate a result in a separate thread while continuing to handle requests on the main thread. The whole point of adding another thread is to avoid timeouts due to calculations that take a long time. But our calculation is very quick. We need to slow the calculation down so that we can test the service with a reasonable response time.

How will we slow down the thread? And what amount of time should the calculation require to complete? These are the questions that we've been building code to answer in this chapter. The next section will explain how you can test services that use multiple threads. And now that we have a service that uses another thread for the calculation, we can explore the best way to test this situation.

I'd also like to clarify that what the next section does is different than adding a delay to the five service call attempts. A delay in the test loop will improve the reliability of the test we have now. The next section will remove the loop completely and show you how to coordinate a test with another thread so that both the test and the thread proceed together.

How to test multiple threads without sleep

Earlier in this chapter, in the *The need to justify multiple threads* section, I mentioned that you should try to do as much work as possible with single threads. We're going to follow this advice now. In the current request handling for the calculate request, the code creates a thread that does a simple calculation, like this:

```
    std::thread calcThread([calcIndex] ()
    {
        calculations[calcIndex].setData(true, 100, 50);
    });
```

Okay, maybe a simple calculation is the wrong way to describe what the thread does. The thread sets the result to a hardcoded value. We know this is temporary code and that we'll need to change the code to multiply the seed value by 10, which is what the tests expect.

Where should the calculation be done? It would be easy to do the calculation in the thread lambda, but that would go against the advice of doing as much work as possible with a single thread.

What we want to do is create a calculation function that the thread can call. This will let us test the calculation function separately without worrying about any threading issues and make sure that the calculation is correct.

And here's the really interesting part: creating a function to do the calculation will help us test the thread management too! How? Because we're going to create two calculation functions.

One function will be the real calculation function, which can be tested independently of any threads. For our project, the real calculation will still be simple and fast. We're not going to try to do a lot of work to slow down the calculation and we're not going to put the thread to sleep either. And we're not going to write a bunch of tests to make sure the calculation is correct. This is just an example of a pattern that you can follow in your projects.

The other function will be a test calculation function and will do some fake calculations designed to match the real calculation result. The test calculation function will also contain some thread management code designed to coordinate the thread's activity. We'll use the thread management code in the test calculation function to slow down the thread so that we can simulate a calculation that takes a long time.

What we're doing is mocking the real calculation with code that is less focused on the calculation and more focused on the thread's behavior. Any test that wants to test the real calculation can use the real calculation function, while any test that wants to test the thread timing and coordination can use the test calculation function.

First, we'll declare the two functions in `Service.h` right before the `Service` class, like this:

```
void normalCalc (int seed, int & progress, int & result);
void testCalc (int seed, int & progress, int & result);
```

You can define your calculation functions in your projects to do whatever you need. Your functions will likely be different. The main point to understand is that they should have the same signature so that the test function can be substituted for the real function.

The `Service` class needs to be changed so that one of these functions can be injected into the service. We'll set up the calculation function in the constructor and use the real function as the default, like this:

```
class Service
{
public:
```

```
    using CalcFunc = void (*) (int, int &, int &);

    Service (CalcFunc f = normalCalc)
    : mCalc(f)
    { }

    void start ();

    ResponseVar handleRequest (std::string const & user,
        std::string const & path,
        RequestVar const & request);

private:
    CalcFunc mCalc;
};
```

The Service class now has a member function pointer that will point to one of the calculation functions. Which one will be called is determined when the Service class is created.

Let's implement the two functions in Service.cpp, like this:

```
void SimpleService::normalCalc (
    int seed, int & progress, int & result)
{
    progress = 100;
    result = seed * 10;
}

void SimpleService::testCalc (
    int seed, int & progress, int & result)
{
    progress = 100;
    result = seed * 10;
}
```

At the moment, both functions are the same. We'll take this one step at a time. Each function just sets progress to 100 and result to seed times 10. We're going to leave the real or normal function as-is. Eventually, we'll change the test function so that it controls the thread.

Now, we can change the calculate request handler in `Service.cpp` so that it uses the calculation function, like this:

```
if (auto const * req = std::get_
    if<CalculateRequest>(&request))
{
    MereMemo::log(debug, User(user), LogPath(path))
        << "Received Calculate request for: "
        << std::to_string(req->mSeed);

    calculations.emplace_back();
    int calcIndex = calculations.size() - 1;
    int seed = req->mSeed;
    std::thread calcThread([this, calcIndex, seed] ()
    {
        int progress;
        int result;
        mCalc(seed, progress, result);
        calculations[calcIndex].setData(true, progress,
                                   result);
    });
    calcThread.detach();
    response = SimpleService::CalculateResponse {
        .mToken = std::to_string(calcIndex)
    };
}
```

In the thread lambda, we call `mCalc` instead of setting `progress` and `result` to hardcoded values. Which calculation function is called depends on which function `mCalc` points to.

If we build and run the test application, we'll see that the tests pass. But there's something wrong with how we're calling `mCalc`. We want to get intermediate progress so that a caller can make status requests and see the progress increasing until the calculation is finally complete. By calling `mCalc` once, we only give the function one chance to do something. We should be calling the `mCalc` function in a loop until `progress` reaches `100` percent. Let's change the lambda code:

```
    std::thread calcThread([this, calcIndex, seed] ()
    {
        int progress {0};
```

```
            int result {0};
            while (true)
            {
                mCalc(seed, progress, result);
                if (progress == 100)
                {
                    calculations[calcIndex].setData(true,
                    progress, result);
                    break;
                }
                else
                {
                    calculations[calcIndex].setData(false,
                    progress, result);
                }
            }
        });
```

This change does not affect the tests because the `mCalc` function currently sets `progress` to `100` on the first call; therefore, the while loop will only run once. We don't want the thread to take too long to run without some synchronization with the tests because we'll never join with the thread. If this was a real project, we would want to use threads from a thread pool and wait for the threads to complete before stopping the service.

Making a change that does not affect the tests is a great way to verify changes. Take small steps instead of trying to do everything in one giant set of changes.

Next, we're going to duplicate the test that generates a result except we will use the test calculation function in the duplicate test. The test will need to be modified slightly so that it can use the test calculation function. But for the most part, the test should remain almost identical. The new test goes in `Message.cpp` and looks like this:

```
TEST_SUITE("Status request to test service generates result",
"Service 2")
{
    std::string user = "123";
    std::string path = "";

    SimpleService::RequestVar calcRequest =
        SimpleService::CalculateRequest {
```

```
            .mSeed = 5
        };
    auto responseVar = gService2.service().handleRequest(
        user, path, calcRequest);
    auto const calcResponse =
        std::get_if<SimpleService::CalculateResponse>
        (&responseVar);
    CONFIRM_TRUE(calcResponse != nullptr);

    SimpleService::RequestVar statusRequest =
        SimpleService::StatusRequest {
            .mToken = calcResponse->mToken
        };
    int result {0};
    for (int i = 0; i < 5; ++i)
    {
        responseVar = gService2.service().handleRequest(
            user, path, statusRequest);
        auto const statusResponse =
            std::get_if<SimpleService::StatusResponse>
            (&responseVar);
        CONFIRM_TRUE(statusResponse != nullptr);

        if (statusResponse->mComplete)
        {
            result = statusResponse->mResult;
            break;
        }
    }
    CONFIRM_THAT(result, Equals(40));
}
```

The only changes are to give the test a different name so that it uses a new test suite called "Service 2", and then use a different global service called gService2. Here, we expect a slightly different result. We'll be changing this test soon so that it will eventually contribute more value than it does now, and we'll be removing the loop that tries to make the request five times. Making these changes in small steps will let us verify that we don't break anything major. And expecting a slightly different result will let us verify that we are using a different calculation function.

To build the project, we need to define gService2, which will use a new setup and teardown class. Add the following code to SetupTeardown.h:

```
class TestServiceSetup
{
public:
    TestServiceSetup ()
    : mService (SimpleService::testCalc)
    { }

    void setup ()
    {
        mService.start ();
    }

    void teardown ()
    {
    }

    SimpleService::Service & service ()
    {
        return mService;
    }

private:
    SimpleService::Service mService;
};

extern MereTDD::TestSuiteSetupAndTeardown<TestServiceSetup>
gService2;
```

The TestServiceSetup class defines a constructor that initializes the mService data member with the testCalc function. The gService2 declaration uses TestServiceSetup. We need to make a small change in SetupTeardown.cpp for gService2, like so:

```
#include "SetupTeardown.h"

MereTDD::TestSuiteSetupAndTeardown<ServiceSetup>
```

```
gService1("Calculation Service", "Service 1");

MereTDD::TestSuiteSetupAndTeardown<TestServiceSetup>
gService2("Calculation Test Service", "Service 2");
```

The `SetupTeardown.cpp` file is short and only needs to define instances of `gService1` and `gService2`.

We need to change the `testCalc` function so that it will multiply by 8 to give an expected result of 40 instead of 50. Here are both calculation functions in `Service.cpp`:

```
void SimpleService::normalCalc (
    int seed, int & progress, int & result)
{
    progress = 100;
    result = seed * 10;
}

void SimpleService::testCalc (
    int seed, int & progress, int & result)
{
    progress = 100;
    result = seed * 8;
}
```

Building and running the test application shows that all the tests pass. We now have two test suites. The output looks like this:

```
Running 2 test suites
--------------- Suite: Service 1
------- Setup: Calculation Service
Passed
------- Test: Calculate request can be sent
Passed
------- Test: Status request generates result
Passed
------- Teardown: Calculation Service
Passed
--------------- Suite: Service 2
```

```
------- Setup: Calculation Test Service
Passed
------- Test: Status request to test service generates result
Passed
------- Teardown: Calculation Test Service
Passed
----------------------------------
Tests passed: 7
Tests failed: 0
```

Here, we introduced a new service that uses a slightly different calculation function and can use both services in the tests. The tests pass with minimal changes. Now, we're ready to make more changes to coordinate the threads. This is a better approach than jumping directly into the thread management code and adding the new service and calculation function.

When following TDD, the process is always the same: get the tests to pass, make small changes to the tests or add new tests, and get the tests to pass again.

The next step will complete this section. We're going to control the speed at which the testCalc function works so that we can make multiple status requests to get a complete result. We'll wait inside the test calculation function so that the test has time to verify that the progress does indeed increase over time until the result is finally calculated once the progress reaches 100%.

Let's start with the test. We're going to signal the calculation thread from within the test thread so that the calculation thread will progress in-step with the test. This is what I meant by testing multiple threads without using sleep. Sleeping within a thread is not a good solution because it's not reliable. You might be able to get a test to pass only to have the same test fail later when the timing changes. The solution you'll learn here can be applied to your testing.

All you need to do is create a test version of part of your code that can be substituted for the real code. In our case, we have a testCalc function that can be substituted for the normalCalc function. Then, you can add one or more *condition variables* to your test and wait on those condition variables from within the test version of your code. A condition variable is a standard and supported way in C++ to let one thread wait until a condition is met before proceeding. The test calculation function will wait on the condition variable. The test will notify the condition variable when it's ready for the calculation to continue. Notifying the condition variable will unblock the waiting calculation thread at exactly the right time so that the test can verify the proper thread behavior. Then, the test will wait until the calculation has been completed before continuing. We'll need to include condition_variable at the top of Service.h, like this:

```
#ifndef SIMPLESERVICE_SERVICE_H
#define SIMPLESERVICE_SERVICE_H
```

```
#include <condition_variable>
#include <string>
#include <variant>
```

Then, we need to declare a mutex, two condition variables, and two bools in `Service.h` so that they can be used by the test calculation function and by the test. Let's declare the mutex, condition variables, and the bools right before the test calculation function, like this:

```
void normalCalc (int seed, int & progress, int & result);

extern std::mutex service2Mutex;
extern std::condition_variable testCalcCV;
extern std::condition_variable testCV;
extern bool testCalcReady;
extern bool testReady;
void testCalc (int seed, int & progress, int & result);
```

Here is the modified test in `Message.cpp`:

```
TEST_SUITE("Status request to test service generates result",
"Service 2")
{
    std::string user = "123";
    std::string path = "";

    SimpleService::RequestVar calcRequest =
        SimpleService::CalculateRequest {
            .mSeed = 5
        };
    auto responseVar = gService2.service().handleRequest(
        user, path, calcRequest);
    auto const calcResponse =
        std::get_if<SimpleService::CalculateResponse>
        (&responseVar);
    CONFIRM_TRUE(calcResponse != nullptr);

    // Make a status request right away before the service
    // is allowed to do any calculations.
    SimpleService::RequestVar statusRequest =
```

```
        SimpleService::StatusRequest {
            .mToken = calcResponse->mToken
        };
    responseVar = gService2.service().handleRequest(
        user, path, statusRequest);
    auto statusResponse =
        std::get_if<SimpleService::StatusResponse>
        (&responseVar);
    CONFIRM_TRUE(statusResponse != nullptr);
    CONFIRM_FALSE(statusResponse->mComplete);
    CONFIRM_THAT(statusResponse->mProgress, Equals(0));
    CONFIRM_THAT(statusResponse->mResult, Equals(0));

    // Notify the service that the test has completed the first
    // confirmation so that the service can proceed with the
    // calculation.
    {
        std::lock_guard<std::mutex>
            lock(SimpleService::service2Mutex);
        SimpleService::testReady = true;
    }
    SimpleService::testCV.notify_one();

    // Now wait until the service has completed the
calculation.
    {
        std::unique_lock<std::mutex>
            lock(SimpleService::service2Mutex);
        SimpleService::testCalcCV.wait(lock, []
        {
            return SimpleService::testCalcReady;
        });
    }

    // Make another status request to get the completed result.
    responseVar = gService2.service().handleRequest(
        user, path, statusRequest);
```

```
        statusResponse =
            std::get_if<SimpleService::StatusResponse>
            (&responseVar);
        CONFIRM_TRUE(statusResponse != nullptr);
        CONFIRM_TRUE(statusResponse->mComplete);
        CONFIRM_THAT(statusResponse->mProgress, Equals(100));
        CONFIRM_THAT(statusResponse->mResult, Equals(40));
}
```

The test is a bit longer than it used to be. We're no longer making status requests in a loop while looking for a completed response. This test takes a more deliberate approach and knows exactly what it expects at each step. The initial calculation request and calculation response are the same. The test knows that the calculation will be paused, so the first status request will return an uncompleted response with zero progress.

After the first status request has been confirmed, the test notifies the calculation thread that it can continue and then the test waits. Once the calculation is complete, the calculation thread will notify the test that the test can continue. At all times, the test and the calculation thread are taking turns, which lets the test confirm each step. There is a small race condition in the test calculation thread that I'll explain after you've seen the code. A race condition is a problem where two or more threads can interfere with each other and the result is not completely predictable.

Let's look at the other half now – the test calculation function. We need to declare the mutex, condition variables, and the bools too. The variables and the test calculation function should look like this:

```
std::mutex SimpleService::service2Mutex;
std::condition_variable SimpleService::testCalcCV;
std::condition_variable SimpleService::testCV;
bool SimpleService::testCalcReady {false};
bool SimpleService::testReady {false};

void SimpleService::testCalc (
    int seed, int & progress, int & result)
{
    // Wait until the test has completed the first status
request.
    {
        std::unique_lock<std::mutex> lock(service2Mutex);
        testCV.wait(lock, []
        {
```

```
            return testReady;
        });
    }


    progress = 100;
    result = seed * 8;


    // Notify the test that the calculation is ready.
    {
        std::lock_guard<std::mutex> lock(service2Mutex);
        testCalcReady = true;
    }
    testCalcCV.notify_one();
}
```

The first thing that the test calculation function does is wait. No calculation progress will be made until the test has a chance to confirm the initial status. Once the test calculation thread is allowed to proceed, it needs to notify the test before returning so that the test can make another status request.

The most important thing to understand about this process is that the test calculation function should be the only code interacting with the test. You shouldn't put any waits or notifications in the main service response handler or even in the lambda that is defined in the response handler. Only the test calculation function that gets swapped out for the real calculation function should have any awareness that a test is being run. In other words, you should put all the waiting and condition variable notifications in testCalc. This is the source of the race condition that I mentioned. When the testCalc function notifies the test thread that the calculation is complete, it's not completely correct. The calculation is only complete when setData finishes updating CalcRecord. However, we don't want to send the notification after calling setData because that would put the notification outside of the testCalc function.

Ideally, we would change the design so that the calculation function is called one additional time after completing the calculation. We could say that this gives the calculation function a chance to clean up any resources used during the calculation. Or maybe we can create another set of functions for cleaning up. One cleanup function could be the normal cleanup, while the other function could be substituted for test cleanup. Either approach would let us notify the test that the calculation has finished, which would eliminate the race condition.

Building and running these tests shows that all the tests continue to pass. We're almost done. We'll leave the race condition as-is because fixing it would only add extra complexity to this explanation. The only remaining task is to fix a problem that I noticed in the log file. I'll explain more about this new problem in the next section.

Fixing one last problem detected with logging

There's a big reason why I choose to build a logging library in *Part 2, Logging Library*, of this book. Logging can be a huge help when debugging known problems. Something that's often overlooked is the benefit that logging provides when looking for bugs that haven't been detected yet.

I'll often look at the log file after running tests to make sure the messages match what I expect. After making the enhancements in the previous section for the thread coordination between the test and the test calculation thread, I noticed something strange in the log file. The log file looks like this:

```
2022-08-27T05:00:50.409 Service is starting.
2022-08-27T05:00:50.410 user="123" Received Calculate request
for: 5
2022-08-27T05:00:50.411 user="123" Received Calculate request
for: 5
2022-08-27T05:00:50.411 user="123" Received Status request for:
1
2022-08-27T05:00:50.411 Service is starting.
2022-08-27T05:00:50.411 Service is starting.
2022-08-27T05:00:50.411 user="123" Received Calculate request
for: 5
2022-08-27T05:00:50.411 user="123" Received Calculate request
for: 5
2022-08-27T05:00:50.411 user="123" Received Status request for:
2
2022-08-27T05:00:50.411 user="123" Received Status request for:
2
2022-08-27T05:00:50.411 user="123" Received Status request for:
2
2022-08-27T05:00:50.411 user="123" Received Status request for:
2
```

I removed the `log_level` and `logpath` tags just to shorten the messages so that you can see the important parts better. The first strange thing that I noticed is that the service was started three times. We only have `gService1` and `gService2`, so the service should only have been started twice.

The first four lines in the log file make sense. We start `gService1` and then run a simple test that requests a calculation and checks that the response is of the proper type. Then, we run another test that makes a status request up to five times while looking for a complete response. The first status request finds the complete response, so no additional status requests are needed. The token for the first status request is `1`.

Line 5 in the log file, which is where the service is started for the second time, is where the log file begins to look strange. We should only need to start the second service, make a single additional request, and then make two status requests. It looks like the log file is getting duplicate messages from line 5 until the end.

After a little debugging and the hint that we're duplicating log messages, I found the problem. When I originally designed the service, I configured the logging in the `Service::start` method. I should have kept the logging configuration in the `main` function. Everything worked until we needed to create and start a second service so that the second service could be configured to use a test calculation function. Well, the second service was also configuring the logging when it started, and it added another file output. The second file's output caused all the log messages to be sent to the log file twice. The solution is simple: we need to configure the logging in `main` like this:

```
#include <MereMemo/Log.h>
#include <MereTDD/Test.h>

#include <iostream>

int main ()
{
    MereMemo::FileOutput appFile("logs");
    MereMemo::addLogOutput(appFile);

    return MereTDD::runTests(std::cout);
}
```

Then, we need to remove the logging configuration from the service `start` method so that it looks like this:

```
void SimpleService::Service::start ()
{
    MereMemo::log(info) << "Service is starting.";
}
```

With these changes, the tests still pass and the log file looks better. Again, I removed some tags to shorten the log message lines. Now, the content of the log file is as follows:

```
2022-08-27T05:35:30.573 Service is starting.
2022-08-27T05:35:30.574 user="123" Received Calculate request
for: 5
2022-08-27T05:35:30.574 user="123" Received Calculate request
```

```
for: 5
2022-08-27T05:35:30.574 user="123" Received Status request for:
1
2022-08-27T05:35:30.574 Service is starting.
2022-08-27T05:35:30.574 user="123" Received Calculate request
for: 5
2022-08-27T05:35:30.574 user="123" Received Status request for:
2
2022-08-27T05:35:30.575 user="123" Received Status request for:
2
```

While the problem ended up being a mistake in how the logging was configured, the point I wanted to make is to remind you to look through the log files periodically and make sure the log messages make sense.

Summary

This is the last chapter of this book and it explained one of the most confusing and difficult aspects of writing software: how to test multiple threads. You'll find a lot of books that explain multi-threading but fewer will give you advice and show you effective ways to test multiple threads.

Because the target customer of this book is a microservices C++ developer who wants to learn how to use TDD to design better software, this chapter tied everything in this book together to explain how to test multi-threaded services.

First, you learned how to use multiple threads in your tests. You need to make sure you handle exceptions inside tests that start additional threads. Exceptions are important because the testing library uses exceptions to handle failed confirmations. You also learned how to use a special helper class to report failed confirmations that arise in additional threads.

Threads must also be considered when writing and using libraries. You saw how to test a library to make sure it's thread-safe.

Finally, you learned how to test multi-threaded services in a fast and reliable manner that avoids putting threads to sleep in an attempt to coordinate the actions of multiple threads. You learned how to refactor your code so that you can test as much as possible in a single-threaded manner and then how to substitute the normal code for special test-aware code that works with a test. You can use this technique any time you need a test and multi-threaded code to work together so that the test can take specific and reliable steps and confirm your expectations along the way.

Congratulations on reaching the end of this book! This chapter visited all the projects we've been working on. We enhanced the unit testing library to help you use multiple threads in your tests. We also made the logging library thread-safe. Finally, we enhanced the service so that it can coordinate multiple threads between the service and the tests. You now have all the skills you'll need to apply TDD to your projects.

Index

A

assert 59
assertions
supporting, with testing library 67-71

B

beta environment 337
bool confirms
equality, confirming 76-82
fixing 76
build failure 36, 37

C

C++
using, to write tests 11-15
char pointers 305
class 6
classic style 288
confirm 59
types, adding 87-93
writing 101, 102
Coordinated Universal Time (UTC) 180
Curiously Recurring Template Pattern (CRTP) 224

current confirmations
issue 288, 289
custom Hamcrest matchers
writing 327-332

D

default tag values
adding 199-212
dependencies
designing with 270-274
dependency injection 271
development environment 337
Disjunctive Normal Form (DNF) 227

E

epsilon 317
error cases
testing 72-74
errors
handling, in teardown 145-149
handling, in test setup 145-149
expect 59
expected failures 43
Extensible Markup Language (XML) 263

F

filtering

enhancing, to allow filtering based on tag relative values 245-260

options, exploring 212-215

first message

confirming 172-178

logging 172-178

floating-point Hamcrest matchers

adding 322-326

floating-point values

comparisons 310-321

confirming 96-101

function 4

functor 6

G

Greenwich Mean Time (GMT) 180

H

Hamcrest matchers

test library, enhancing to support 294-300

Hamcrest style 289

Hamcrest types

adding 300-307

I

integrated development environment (IDE) 171

L

lambdas

exploring, for tests 108-112

line numbers

obtaining, without macros 106, 107

test failures, decoupling from 82- 87

logging issues

fixing 398-400

logging library

building 164

building, with TDD 165

designing 166-169

making thread-safe 361-367

log information

controlling 235-244

log levels

adding 190-198

log messages

constructing, with streams 181-184

filtering, with tests 226- 234

M

macros

not using, for obtaining line numbers 106, 107

magic numbers 68

microservices 336

missed failure 54

counting 56

monolithic application 336

multiple logging outputs

adding 274- 284

multiple service calls

making 379-385

multiple tests

supporting, with test declaration 25-28

multiple threads

testing, without using sleep 386- 397

with code, testing 367-375

N

namespace 11

O

output results
redirecting 31, 32

P

policy 121
policy class 121
production environment 337
project
testing 159

R

refactoring 82
requires feature 308
results
summarizing 28-31

S

service 338
testing, considerations 338, 339
service path 345
service return type
changing 375-379
service testing
challenges 336-338
setup 115
SimpleService project 340-350
single test result
reporting 20-25

streams
log messages, constructing with 181-184
string confirmations
simplifying 290-294
string literals
confirming 94-96

T

tag design
refactoring, with TDD 221-226
tag types
adding 216-220
teardown 115
enhancing, for multiple tests 123-145
errors, handling 145-149
supporting 116-123
test 4
appearance 5, 6
designing, by filtering log messages 226-234
easy understandability 152, 153
enhancement, for obtaining
another pass 43-57
features 151
gaps, finding 188, 189
improving, with descriptive names 152, 153
information requirements 6-10
internal steps, checking 159
lambdas, exploring for 108-112
multiple threads, using 352-361
pass or failure, detecting 60-66
random behavior, using 157, 158
specific scenario focused tests 154-157
types 265, 266
using 15-18
writing, to pass 38-43
writing, with C++ 11-15

test declaration

enhancing, to support multiple tests 25-28

test-driven development (TDD) 310

integration test 264, 265

logging library, building with 165

used, for starting project 169-171

system tests 264, 265

tag design, refactoring with 221-226

test failures

decoupling, from line numbers 82-87

testing

limitation 261-263

testing library

enhancing, to support assertions 67- 71

enhancing, to support Hamcrest
matchers 294-300

test setup

enhancing, for multiple tests 123-145

errors, handling 145-149

supporting 116-123

timestamps

adding 179, 180

Packt.com

Subscribe to our online digital library for full access to over 7,000 books and videos, as well as industry leading tools to help you plan your personal development and advance your career. For more information, please visit our website.

Why subscribe?

- Spend less time learning and more time coding with practical eBooks and Videos from over 4,000 industry professionals
- Improve your learning with Skill Plans built especially for you
- Get a free eBook or video every month
- Fully searchable for easy access to vital information
- Copy and paste, print, and bookmark content

Did you know that Packt offers eBook versions of every book published, with PDF and ePub files available? You can upgrade to the eBook version at packt.com and as a print book customer, you are entitled to a discount on the eBook copy. Get in touch with us at customercare@packtpub.com for more details.

At www.packt.com, you can also read a collection of free technical articles, sign up for a range of free newsletters, and receive exclusive discounts and offers on Packt books and eBooks.

Other Books You May Enjoy

If you enjoyed this book, you may be interested in these other books by Packt:

C++20 STL Cookbook

Bill Weinman

ISBN: 9781803248714

- Understand the new language features and the problems they can solve
- Implement generic features of the STL with practical examples
- Understand standard support classes for concurrency and synchronization
- Perform efficient memory management using the STL
- Implement seamless formatting using std::format
- Work with strings the STL way instead of handcrafting C-style code

Template Metaprogramming with C++

Marius Bancila

ISBN: 9781803243450

- Understand the syntax for all types of templates
- Discover how specialization and instantiation works
- Get to grips with template argument deduction and forwarding references
- Write variadic templates with ease
- Become familiar with type traits and conditional compilation
- Restrict template arguments in C++20 with constraints and concepts
- Implement patterns such as CRTP, mixins, and tag dispatching

Packt is searching for authors like you

If you're interested in becoming an author for Packt, please visit `authors.packtpub.com` and apply today. We have worked with thousands of developers and tech professionals, just like you, to help them share their insight with the global tech community. You can make a general application, apply for a specific hot topic that we are recruiting an author for, or submit your own idea.

Share Your Thoughts

Now you've finished *Test-Driven Development with C++*, we'd love to hear your thoughts! Scan the QR code below to go straight to the Amazon review page for this book and share your feedback or leave a review on the site that you purchased it from.

`https://packt.link/r/1803242000`

Your review is important to us and the tech community and will help us make sure we're delivering excellent quality content.

Download a free PDF copy of this book

Thanks for purchasing this book!

Do you like to read on the go but are unable to carry your print books everywhere? Is your eBook purchase not compatible with the device of your choice?

Don't worry, now with every Packt book you get a DRM-free PDF version of that book at no cost.

Read anywhere, any place, on any device. Search, copy, and paste code from your favorite technical books directly into your application.

The perks don't stop there, you can get exclusive access to discounts, newsletters, and great free content in your inbox daily

Follow these simple steps to get the benefits:

1. Scan the QR code or visit the link below

https://packt.link/free-ebook/9781803242002

2. Submit your proof of purchase
3. That's it! We'll send your free PDF and other benefits to your email directly

www.ingramcontent.com/pod-product-compliance
Lightning Source LLC
Chambersburg PA
CBHW081501050326
40690CB00015B/2880